P9-CQG-277

dawn of man

THE STORY OF HUMAN EVOLUTION

dawnofman

THE STORY OF HUMAN EVOLUTION

ROBIN McKIE

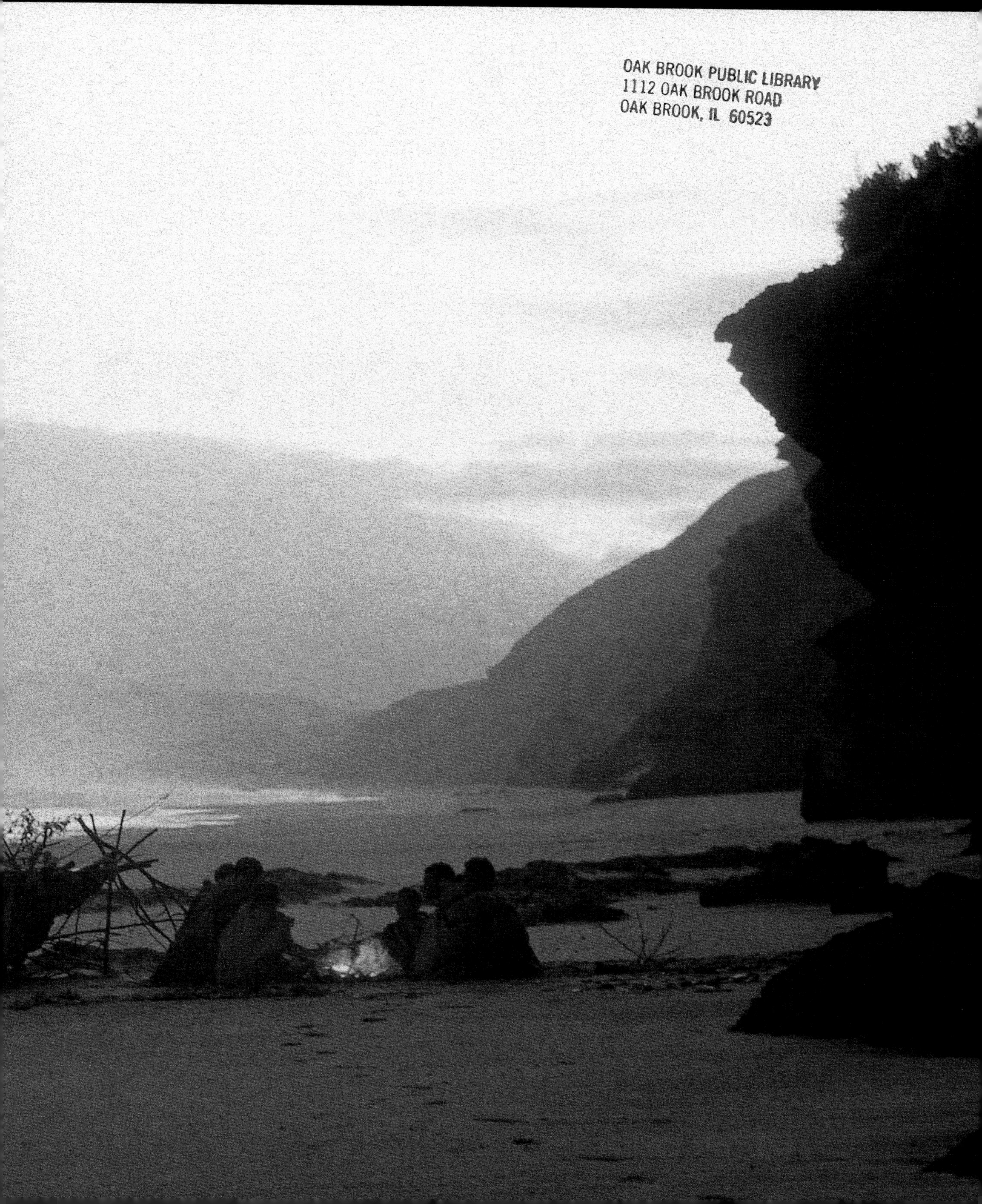

This book is published to accompany the BBC television series *ape•man*, first
broadcast in the United States as *Dawn of Man* on The Learning Channel in 2000.
Executive producer: John Lynch
Series producer: Philip Martin
Producers: Harvey Jones, Jeff Morgan,
Lisa Silcock, Charlie Smith and David Wilson

Commissioning editor: **Sheila Ableman**
Scientific consultant: **Leslie Aiello**
Project editor: **Lara Speicher**
Text editor: **Ben Morgan**
Designer: **John Calvert**
Picture researcher: **Miriam Hyman**
Illustrator: **James Robins**
Maps: **Olive Pearson**

First published in 2000 by BBC Worldwide Ltd,
80 Wood Lane, London W12 0TT

Publisher: **Sean Moore**
Editors: **Jill Hamilton, Barbara Minton**
Art director: **Dirk Kaufman**

First American Edition, 2000
10 9 8 7 6 5 4 3 2 1
Published in the United States by Dorling Kindersley Publishing, Inc.
95 Madison Avenue, New York, New York 10016

ISBN 0-7894-6262-1

Set in Gill Sans
Printed and bound in Great Britain by Butler & Tanner Ltd., Frome and London
Color separations by Radstock Reproductions Ltd, Midsomer Norton
Jacket printed by Lawrence Alllen Ltd., Weston-Super-Mare.

CONTENTS

PREFACE

Human beings have existed on this planet for an absurdly short time. Yet we now dominate its landscapes to such an extent that we have become, whether we like it or not, Earth's stewards. The fates of hundreds of species are likely to be determined by our actions over the next few decades, a power that seems to set us apart from other animals, and which appears to confirm the widespread belief that *Homo sapiens* is a special, exalted, godlike creature. We are nothing of the kind, of course, as an appreciation of our origins – a primate that left the trees of Africa a mere 5 million years ago – makes clear. In arriving at its current state, our species was driven by the same evolutionary forces that shaped all other living beings. Unraveling the details of this process is now one of the most important services that science can provide for mankind.

By good chance, it is also one of science's most dramatic stories and is the handiwork of a group of researchers – working in a series of seemingly disparate professions – who have combined their talents to revolutionize our self-image. These individuals include paleontologists, who study the bones of long-dead animals; anthropologists, who research human characteristics and beliefs; archaeologists, who analyze ancient man-made objects; and molecular biologists, who use the genetic makeup of men and women today to reveal precious data about past populations. These are the true heroes of this book: men and women whose endeavors – carried out in sweltering heat, in cramped, claustrophobic caverns, or in arid deserts – are transforming our understanding of ourselves.

It is also a story riven with jealousies and disagreements, like so many other human endeavors. Only a handful of our ancestors' fossils have ever been found. Yet scientists are often obliged to interpret them in ways that have profound implications for mankind today. Are we bloodthirsty, atavistic killers at heart, or are we descended from tribes of happy vegetarians? A simple find can transform one career and the analysis of its significance ruin another. Not surprisingly, mistakes litter the path of paleontology, as in the case of the Piltdown hoax, which, as we shall see, misled a generation of scientists about the course of human evolution. And, no doubt, errors will continue to be made, though this is only likely to make our story even more dramatic, mysterious, and exciting – if such a thing were possible.

This, in short, is one of the greatest tales ever told – an epic that begins in Africa 5 million years ago with the first appearance of our apelike ancestors and ends with the emergence of *Homo sapiens* and its art, culture, and technology. It has been a privilege to have been involved in this great saga and to have been involved in the writing of one version of that story. For that I have to thank many people who provided crucial assistance. I am particularly indebted to Leslie Aiello, who read my manuscript and offered invaluable advice and information. Meave Leakey also checked several chapters and provided key insights into her family and work, while many other scientists provided details and advice about key components of my narrative. They include Paul Abell, Peter Andrews, Juan Luis Arsuaga, Clive Gamble, Andrew Hill, Ann MacLarnon, Richard Potts, Mark Roberts, Chris Stringer, Carl Swisher, Erik Trinkaus, Alan Walker, and Tim White. I would also like to thank my editor, Lara Speicher, and Sheila Ableman of BBC Worldwide, for their encouragement and guidance, and also Philip Martin and Lisa Silcock of BBC Television for their expertise and counsel. However, most of all, I would like to thank my wife, Sarah, who has provided precious help, patience, and support and who has shared my enthusiasm for this, the most exciting of all scientific odysseys.

Robin McKie
November 1999

CHAPTER ONE

OUT OF THE WOODS

There is a host of complex aids at the disposal of the modern fossil hunter, but there is only one that is truly indispensable – and that is luck. Without an occasional dash of good fortune, even the most hard-working paleontologist is doomed to failure, as most will cheerfully admit. Many of the greatest discoveries made about our prehistory have occurred when scientists were just about to close down a dig, or when a researcher was wandering around a site relaxing after work. We shall come across several examples of such scientific serendipity in this book. However, even by normal fortuitous standards, providence was unusually kind to paleontologist Andrew Hill during one excavation in Africa – when he was forced to duck to avoid being struck by a lump of elephant dung. It was a bizarre maneuver to say the least, yet it led to one of the greatest discoveries in the study of human evolution: the uncovering of the Laetoli footprints. These imprints, made by ancient apemen on a stretch of wet ash in northern Tanzania, provide us with one of the most potent images that we have of our evolutionary roots. They show that, more than 3.5 million years ago, our distant kin were already traveling around as we do today: upright and striding. These 54 steps – engraved on a field of damp pumice eons ago – demonstrate that we are, first and foremost, a two-legged mammal. We may take the business of walking on two feet – bipedalism – for granted. Yet it defines us in a remarkable way. Only when we had become an upstanding species could we evolve into the kind of creatures that we are today, and the fossilized evidence of this upright gait, enshrined at Laetoli, provides solid testimony to that fact.

These footprints in the cinders of time were discovered during an expedition in 1976 by the late Mary Leakey, matriarch of the world's most remarkable fossil-hunting clan: a family that included her husband Louis, her son Richard, and her daughter-in-law Meave. All have played critical roles in uncovering parts of our great evolutionary story, as we shall see. However, Mary Leakey's work at Laetoli is perhaps the most striking.

In 1976, her team had been working at Laetoli (the Masai word for the local red lily[1]) for two seasons and had scoured the sediments for fossils of ancient human ancestors, accruing a respectable collection of teeth and bone fragments in the process. (Teeth tend to be the most abundant fossils found at sites because they contain enamel, which is the hardest and therefore the most enduring substance in the body.) It was exhausting work, of course, and one evening, Hill and his colleagues were relaxing by pelting each other with dry elephant droppings – an odd sort of pastime perhaps, but then such are the pressures of modern paleontology. "We had been working really hard that day and were heading back toward camp when one of our team decided to liven things up by slinging elephant dung at the rest of us," recalls Hill. "He aimed one at me, and I had to dive out of the way. I ended up flat on my face. I started to rise and saw marks in the ground. I realized they were fossilized raindrops. Then I looked around and saw ancient animal footprints all over the place. We had passed over that ground so many times before that evening, but none of us had noticed a thing. But once we saw the first prints, we could see them everywhere: fossilized tracks of rhinos, elephants, antelopes, all sorts of animals."[2] The prints, it turned out, had been made in volcanic tuff, a type of sediment created by ash thrown up during eruptions, and this had preserved them, unseen, for eons – until Leakey and her dung-hurling colleagues stumbled on them. "As is so often the case in pivotal discoveries, luck intervened", she later recalled.[3]

Mary Leakey at work on the Laetoli footprints in 1979. She considered this discovery to be her greatest achievement.

Garniss Curtis, a scientist at the University of California, Berkeley, examined the sediments and found that the layer below the prints was rich in large crystals of black mica, which he was able to show were 3.8 million years old, while the level above the tracks was found to be 3.6 million years old.[4] The prints had therefore been made, and preserved, sometime within that span. This geologic sandwich, Leakey's scientists realized, was the handiwork of the nearby volcano of Sadiman, which had belched out ash that had the consistency of fine beach sand. The ash was rich in carbonatite, a substance that acts like cement when wet. Not long after Sadiman erupted, rain turned the blanket of ash into a thin paste. Animals wandered across this delicate matrix, leaving behind perfect casts of their paws, hooves, and feet. Then the sun came out and dried the footprints before Sadiman erupted again, depositing another layer of ash. The tracks were buried, until they were exposed to human gaze after more than 3.5 million years of erosion.[5]

On finding the tracks, Leakey's team abandoned their hunt for more fossils and, for the next two years, continued to excavate at Laetoli, plotting the movements of long-dead hares, guinea fowl, elephant, giraffe, saber-toothed cat, an extinct species of horse called a hipparion, and an odd, long-legged, clawed herbivore – also extinct – known as a chalicothere.[6] Leakey wanted some of these prints removed and exhibited

at a small museum that she ran at Olduvai, not far from Laetoli. She asked geochemist Dr Paul Abell, of Rhode Island University, to quarry out a block of rhino tracks. "It took a few days to cut them and move them. Then I decided to have a stroll around to see if I could find any new animal tracks," Abell recalls. "I came across a stretch of ground where the volcanic tuff had thinned away, exposing a long strip of ash underneath – and on one section I could see the imprint of a human heel. I went around the other side of the exposed section to see if the trail appeared there, and looked back. One of our geologists was about to smash his hammer into the footprint. He was looking for interesting rock formations and all he could see was a volcanic ash deposition. I was looking for animal tracks, and that is all I could see. Fortunately, I managed to stop him."[7]

The ground around the print was carefully cleared, exposing two sets of tracks that trailed off to the north, looking for all the world like the marks left by a couple of vacationers strolling on a wet beach. In some of the prints, scientists could make out the shape of the arch of the foot, the big toe, even the ball of the foot. This was a species adept at walking upright. You could see where the heel struck the ground as a foot came down, followed by a push-off from the toes as the animal took its next stride. And, crucially, the big toes did not splay out from the rest of the foot as they do in other primates, which use all four limbs for climbing and grasping. These creatures were expert walkers. It was a stunning discovery. As Leakey later put it, "Two individuals, one larger, one smaller, had passed this way, 3.6 million years ago."[8]

But who were these individuals? Mary Leakey, who died in 1996 and who regarded Laetoli as the crowning achievement of her six decades of work in East Africa, always refused to assign a species to the track-makers[9] – although she did believe that they stood in the direct line of human ancestry. Most other paleontologists are convinced about their identity, however. They maintain that the print-makers were members of the species *Australopithecus afarensis*. These apemen lived in Africa between 3 and 4 million years ago, though we would be hard pushed to recognize them as our relatives. They were less than 5 ft (1.5 m) high, possessed brains little bigger than those of apes, and had virtually no foreheads. Their arms were long, their legs short, their chests pyramid-shaped like apes, their muzzles pronounced, and their large molar teeth were suitable for chewing nuts, berries, and seeds. This was not Burt Lancaster and Deborah Kerr holding hands as they wandered along the beach, in other words. However, as paleontologist Ian Tattersall, curator of physical anthropology at the American Museum of Natural History, says: "The Laetoli footprints are astonishingly similar to those that barefoot modern people make on a muddy path. They are so similar, in fact, that some scientists have difficulty in believing that they were made by the same early humans whose bones were found at Laetoli."[10] Nevertheless, several *afarensis* fossils have been found in sediment beds around Laetoli, and the species is generally accepted as having made those ancient tracks.

What is particularly striking about Laetoli is the way this ancient trail allows us to follow the actions of long-dead individuals. "Usually behavior has to be inferred indirectly from the evidence of bones and teeth, and there is almost always argument over inferences of this kind," adds Tattersall. "But at Laetoli, through these footprints, behavior itself is fossilized."[11] For one thing, the two sets of prints are extremely close together: either one ape-person was following the other slightly to one side, or perhaps they were in physical contact. It is possible that the larger one was a male, and he could have been placing a protective arm around the smaller female, although this interpretation, displayed in

The Laetoli footprints (opposite) laid down 3.6 million years ago by bipedal apemen, probably *Australopithecus afarensis*, as depicted (below) at the American Museum of Natural History, New York.

STEPPING OUT

Human beings are not the only primates to move around on two legs. Chimpanzees and gibbons also stand upright and walk – but only occasionally. By contrast, *Homo sapiens* is the only primate species that is habitually bipedal. When we walk, in the same way as our australopithecine ancestors did, we push one leg off the ground, putting most pressure on the big toe. The leg then swings under our body in a slightly flexed position and then, as it extends itself, strikes the ground, with the heel hitting first. This leg then provides support for the body as the other leg begins its swing and strike motion. The impact of the balls of the heels and the big toes of *afarensis* folk as they strode over the hot ash of Sadiman volcano provide stark testimony to their bipedalism.

Several critical anatomical changes were required before our ancestors could achieve such upright prowess, however. For one thing, they had to be able to extend their knee and hip joints to create a straight leg, thus reducing the amount of muscle power needed to support their bodies. A chimpanzee cannot do this, and can only support its weight by using its leg muscles, a very tiring business. Our ancestors also had to make sure their centers of gravity remained underneath their bodies so that they did not wobble around as they moved. To achieve this, the human femur had to become angled beneath the pelvis so that our feet could be placed under our centers of gravity as we walked. An ape's unangled femur means it can only waddle when it moves upright. In addition, we had to evolve a curved lower spine and an enlarged big toe which was brought into line with the other toes.

KNEE JOINTS AND PELVISES COMPARED

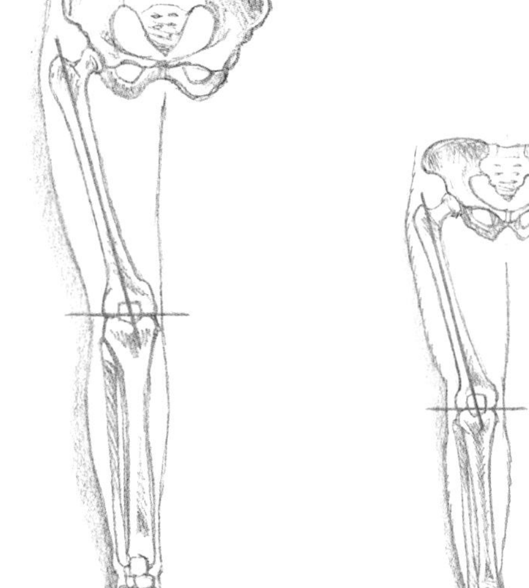

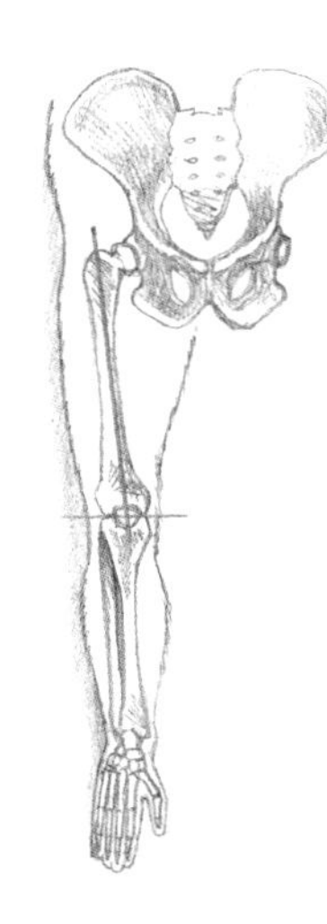

Human (far left)
The femur (thigh bone) of a human being slopes inward from the thigh to the knee so that the feet are directly underneath the body's center of gravity, thus preventing it from moving about very much when a person takes a stride.

Australopithecine (center)
This is also angled rather like a modern human being's, suggesting that *Australopithecus afarensis* was already a better perambulator than apes.

Ape (near left)
The thigh bone of an ape does not slope inward toward the knee and so creates the distinctive waddling upright gait of creatures like the chimpanzee.

COMPARISON OF APE AND HUMAN FEET

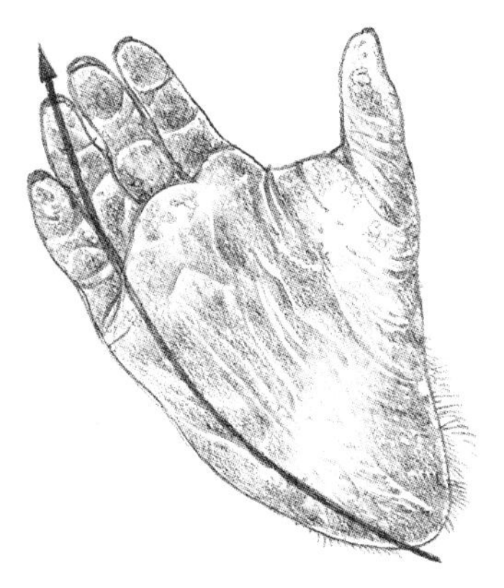

When a chimpanzee walks on two feet, its body weight is placed on the side of its feet. When it takes a step, the animal pushes off from a point in the middle of its row of toes.

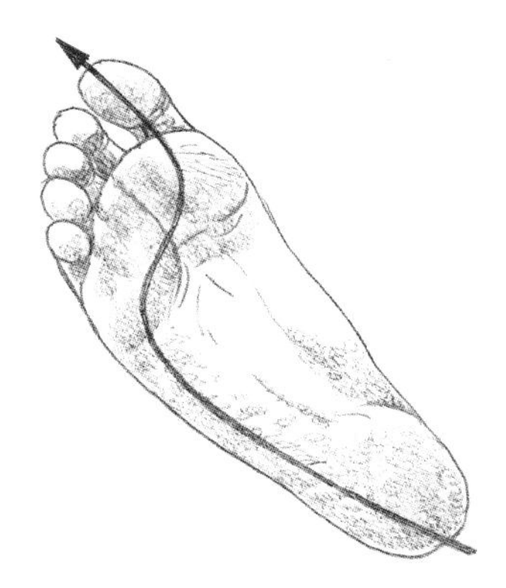

When a man or a woman walks, the weight is transmitted along the outside of the foot near the heel, but is transferred across the ball of the foot and onto the big toe from which a person pushes off when walking, as can be seen in one of the actual Laetoli footprints shown left. This is a more efficient arrangement.

A recreation of a female australopithecine with her baby. Groups of these early apemen survived climatic changes, attacks from predators and the harsh conditions of Africa's savanna.

a diorama in the Hall of Human Biology at Tattersall's museum, has infuriated some feminists as being "paternalistic." For his part, Tattersall defends the interpretation as having "the fewest unwanted implications."[12]

At the time the Laetoli steps were being laid down, this area of East Africa was made up of savanna interspersed with clumps of acacia trees. Sadiman was probably rumbling in the distance and our *afarensis* protagonists may well have been walking from the cover of woods into the open, a maneuver replete with danger. Indeed, at one point, the smaller individual, which Leakey also presumed to be female, clearly halts in her stride. "Following her path produces, at least for me, a kind of poignant wrench. At one point, and you need not be an expert tracker to discern this, she stops, pauses, turns to the left to glance at some possible threat or irregularity, and then continues to the north. This motion, so intensely human, transcends time. Three million, six hundred thousand years ago, a remote ancestor – just as you or I – experienced a moment of doubt."[13]

However, the feature that has probably caused most surprise is one that is very hard for the untrained eye to spot. Scientists noticed that, while the footprints of the smaller individual stand out clearly, those of the other are blurred, "as if he shuffled or dragged his feet," as Leakey once put it.[14] It is also possible that the ash was drier and dustier where he walked, so his imprints formed less clearly. However, Leakey later concluded that the best explanation was that the larger, blurred tracks may actually be double footprints. In other words, a third ape-person followed in the footsteps of the bigger australopithecine, smudging his trail in the process – "perhaps to make it easier to cross the slick, ash-covered ground," as Neville Agnew and Martha Demas – leaders of the Getty Conservation Institute, which has been responsible for preserving and protecting the Laetoli steps – state in *Scientific American*.[15] This idea is backed by Abell. "The material thrown out by Sadiman is rich in sodium carbonate, as well as calcium carbonate, and becomes caustic when wet," he says. "It would have taken the skin off anyone who had walked on that ash for any length of time. Maybe the third person was just trying to protect their feet."[16] Or perhaps Leakey had found, preserved in ancient ash, that most human of actions – a child skipping behind a parent, placing its feet exactly where its elders had walked (although it would have to have been a fairly large boy or girl to make such long strides).

Laetoli is a double treasure.[17] It brings poignancy to the lives of our ancient hominid forebears (a hominid is the name given to any upright-walking primate). However, it also demonstrates – in a way that cannot be matched by any archaic leg bone or pelvis – that we had already made the first critical move on our haphazard journey from tree-loving apes to creatures who would travel to the Moon and who would probe the mystery of the Universe's origins: we were standing and walking on two feet. "One cannot overemphasize the role of bipedalism in hominid development," Leakey stated in a *National Geographic* article that she wrote about Laetoli. "It stands as perhaps the salient point that differentiates the forebears of man from other primates. This unique ability freed the hands for myriad possibilities – carrying, tool-making, intricate manipulations. From this single development, in fact, stems all modern

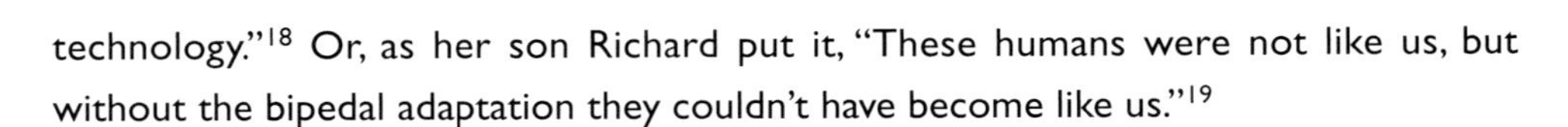

technology."[18] Or, as her son Richard put it, "These humans were not like us, but without the bipedal adaptation they couldn't have become like us."[19]

Now it is usually accepted that three key attributes define human beings: our ability to walk upright, our ability to make tools, and our large brains. But which came first? For much of the earlier part of this century, it was assumed that brain enlargement was the first of these critical evolutionary developments. Our brains got bigger and so created a need to free our hands and arms so that we could make the tools that our developing intellects were bursting to invent. Not so. It is now clear from fossil finds, and from Laetoli, that upright stance came first; brains and tools came later – much later, as it turns out. In other words, the australopithecines who left their mark at Laetoli had made only the first crucial move that would lead to the evolution of *Homo sapiens* (which means "wise man"). They were not really human, in fact. For one thing, their brains – as we have mentioned – were only the size of apes' brains: 350–500 cc in volume, compared with modern humans' average of 1200–1600 cc. Equally, there are no signs that *afarensis* was a tool-maker. (One should note that this involves more than merely being a tool-user, for a Californian sea otter sometimes uses a stone anvil to crack open shellfish, several species of Darwin's finches, from the Galápagos Islands, use cactus spines and twigs to dig for insects, and the Australian black-breasted buzzard is noted for carrying rocks and for dropping them on emu eggs to break them open. Making a tool for some future task implies its creator is capable of a fairly advanced degree of planning.) Not a single stone implement has ever been found at the Laetoli sediment beds, nor has any such utensil been discovered at any other *afarensis* site. "With their hands free, one would have expected this species to have developed tools or weapons of some kind," Mary Leakey reported. "But, except for the ejecta of erupting volcanoes, we haven't found a single stone introduced into the beds. The hominids we discovered had not yet attained the tool-making stage."[20]

However, they had already undergone the most important, and the most difficult, evolutionary change that our ancestors would ever make – and in a remarkably short period of time. In adopting an upright stance, early hominids had to go through an entire suite of anatomical changes, and these changes had already become well developed by the arrival of *Australopithecus afarensis*. By comparing the blood proteins and genes of humans and chimps, scientists know that the lineage that would lead to *Homo sapiens*, and the line that would end up as the common chimpanzee and its close sibling the bonobo, or pygmy chimpanzee, must have parted company about

5 or 6 million years ago. By 3.6 million years before present, and almost certainly much earlier than this (as we shall see), virtually all the differences that separate the human frame from that of apes were evident – in some form – in the australopithecine skeleton. We shall discuss the makeup of these remarkable creatures later in this chapter, but first we should contemplate just what a mighty biological feat it is to be a bipedal primate.

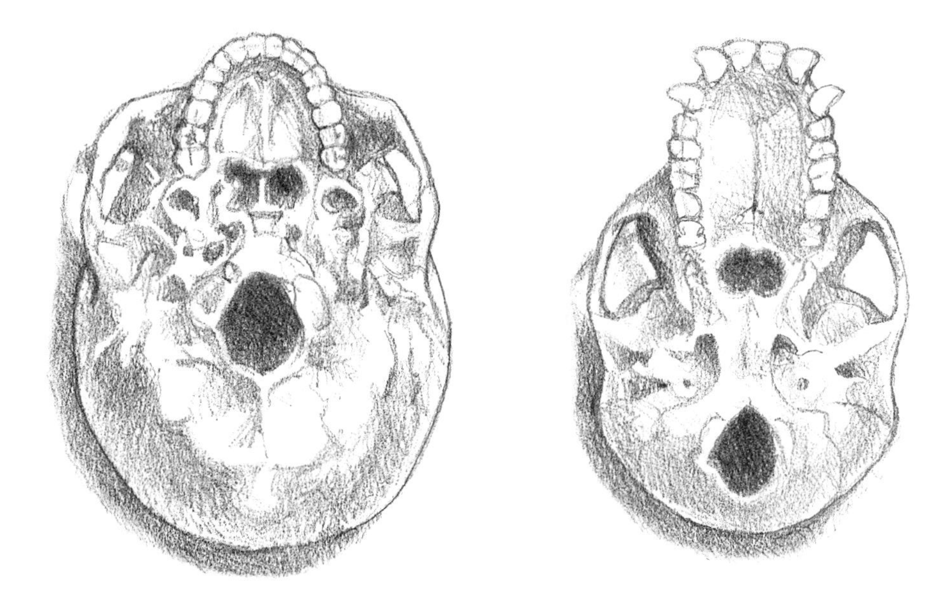

The foramen magnum of a human being (above left) shows the entry of the spinal cord in the center of the skull, the hallmark of an animal that walks upright. On the right, the skull of a chimpanzee – the entry of the spinal cord further back into the skull indicates that the head is not balanced on a vertical spine and the animal does not walk upright.

Consider your knees, for example. They can lock to produce a straight leg that can support your weight without your exerting muscular power. Chimpanzees cannot do this. They do occasionally walk erect, but such a posture is much more tiring for them. In addition, the human femur – the thighbone – slopes inward toward the knee, which prevents us from having an inefficient waddling gait. In humans, the hole at the bottom of the skull through which the spinal cord passes – the foramen magnum – is located in the center of the base of the skull.[21] In apes, the foramen magnum is at the back, allowing them to face forward easily when walking on all fours. And our big toes have lost the thumblike appearance that is typical of apes and have evolved to lie in line with the rest of our toes. *Afarensis* had evolved these features, although not fully. It is an impressive, not to say startling, set of adaptations, as paleontologist Stephen Jay Gould of Harvard University acknowledges. "Upright posture is the surprise, the difficult event, the rapid and fundamental reconstruction of our anatomy," he says.

COMMON ANCESTOR

Human beings share, to the nearest approximation, 98 percent of their DNA with chimpanzees. By contrast, the latter share only 97 percent of their genetic material with gorillas – which means that humans are actually more chimpanzee-like than gorillas are, as the science writer Matt Ridley has pointed out. Scientists believe that about 7 million years ago gorillas diverged from a lineage that split a little later, about 5 or 6 million years ago, into two lines: one that led to modern humans, the other to pygmy and common chimpanzees. But what exactly was this original lineage of "common ancestors" like?

Answering that question has proved to be one of the trickiest in modern paleontology and is very far from being resolved, despite great efforts by scientists. Apes flourished, and spread from Africa into Asia, between 20 to 30 million years ago, and then began a slow decline in the face of changes in the climate, which became cooler and drier. Only the lineages that led to gibbons and orangutans survived in Asia. In Africa, they evolved into humans, chimps, and gorillas. However, there is no fossil record for the last two on this list, making it very hard to guess the nature of our common ancestor.

The possible candidates that have been considered include creatures called Sivapithecus, Ouranopithecus, Rudapithecus, and Dryopithecus, which are dated at about 10–12 million years old, as well as an ape known as Proconsul, which lived about 16–20 million years ago. However, none of these early apes displays all of the characteristics that we would expect in a common ancestor of humans, chimps, and gorillas. For example, those having skeletons that might qualify them as the ancestor don't have the right kind of teeth, and those having the right kind of teeth and jaws have the wrong kind of skeleton – and so the mystery remains.

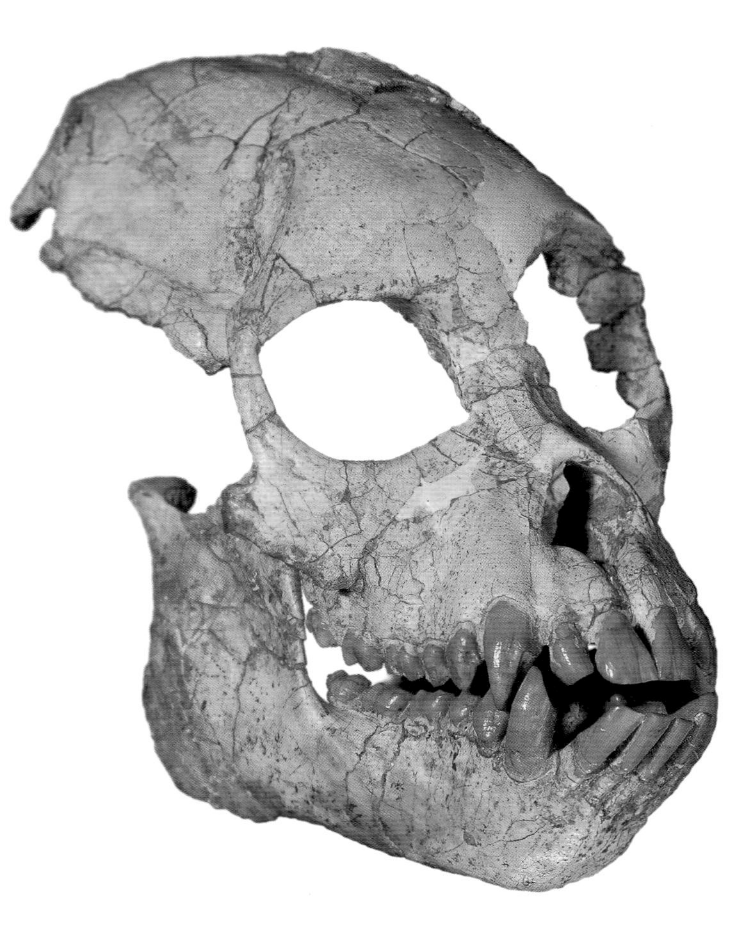

An 18-million-year-old *Proconsul* skull.

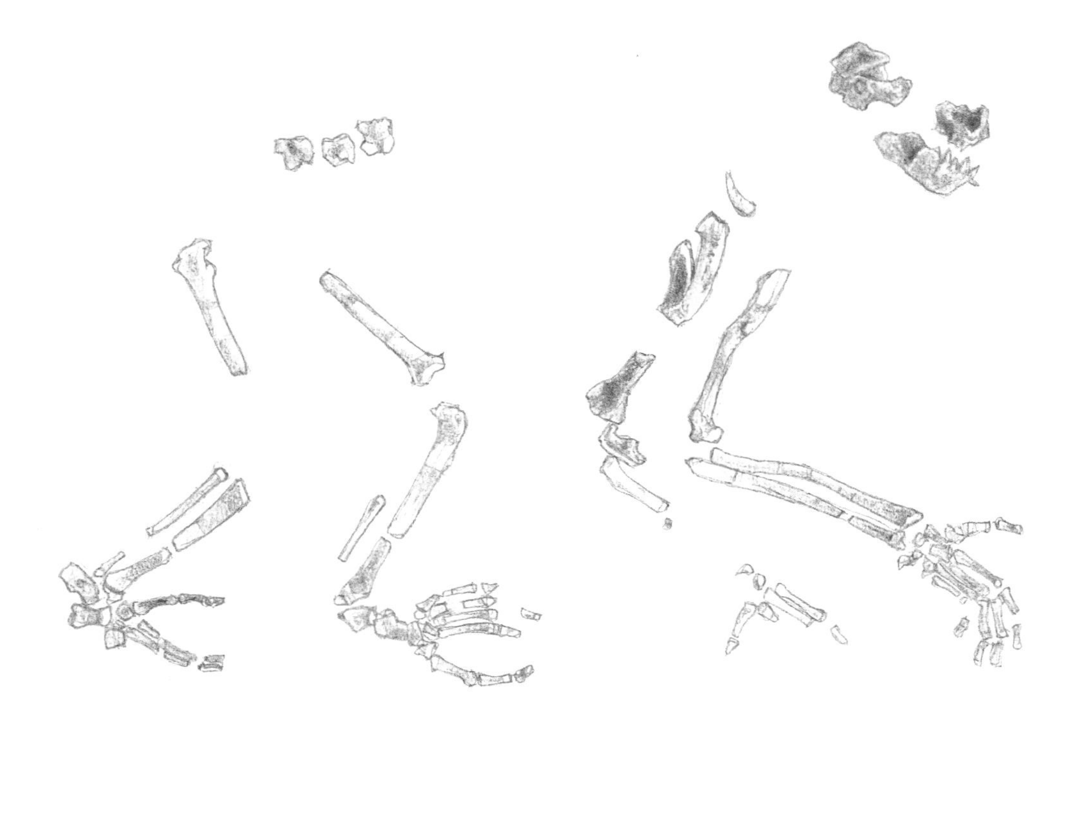

Proconsul – a possible early ancestor of apes and man. A *Proconsul* skeleton (illustrated left) was disovered in 1948 on Rusinga Island in Lake Victoria, Kenya, by the ubiquitous Mary Leakey. The artist's impression below shows how *Proconsul* might have appeared.

"The subsequent enlargement of our brain is, in anatomical terms, a secondary epiphenomenon, an easy transformation embedded in a general pattern of human evolution. As a pure problem in architectural reconstruction, upright posture is far-reaching and fundamental, an enlarged brain superficial and secondary. But the effect of our large brain has far outstripped the relative ease of its construction."[22]

Nor has this surprising catalog of changes been achieved without paying a price. Walking on two legs produces greater wear and tear on our hips, which have to bear our entire body weight. In other primates, this load is shared over four limbs. The consequence of being bipedal humans is disablement – in our later years – that can only be put right through hip-replacement operations. Nor is it necessarily more efficient to get around on two legs as opposed to four. When running, a four-legged primate expends less energy than a bipedal one, although walking on two legs is more efficient than on four. As Owen Lovejoy, an anatomist at Kent State University in the US, observes: "The move to bipedalism is one of the most striking shifts in anatomy you can see in evolutionary biology."[23]

A recreation of an australopithecine mother and child. As forest areas dwindled due to extreme shifts in climate, hominids like these were forced onto open savanna in search of food.

So why did we start walking upright in the first place? What changes had been wrought in East Africa – whose fossil riches indicate this must have been the birthplace of the human lineage – that led one primate species to abandon a four-limbed gait for one involving only two legs? And when exactly did we begin this momentous conversion? In every case, the answer comes down to changes in the environment, the key process that drives evolution. Five million years ago, the world's climate began to deteriorate, going through episodes of drying and seasonal bouts of rainfall. This would have reduced forests and increased grassy woodlands like the savanna, interspersed with acacia, at Laetoli. Inexorably, the habitat of the apes – the common ancestors we share with the chimpanzee – was cut back, forcing those who lived in the eastern part of the continent to begin to forage into open country for at least part of their lives. Bipedalism evolved as a consequence, although exactly why is open to debate. Did our ancestors stand, and stay, upright so that they could keep an eye out for predators on the savanna? Well, possibly, but if they did, it might have got them into more trouble rather than less, as paleontologist Donald Johanson, of the University of Arizona, points out. "It is difficult to imagine a more vulnerable time to try out a major behavioral change than when moving into an unfamiliar habitat, especially one containing successful predators. Imagine how easily large cats could have taken a slow, lumbering hominid that announced its presence by standing up."[24] Or perhaps they walked on two legs to free their hands to carry food back to home bases, or nests, which they still made in the high, safe branches of the trees of those shrinking forests.

Alternatively, there is the ingenious idea put forward by Peter Wheeler of John Moores University in Liverpool. He believes our predecessors adopted an upright stance to minimize the amount of skin that was exposed to the sun's harsh rays when they started to walk away from the shade of their old forest homelands. This prevented their brains from overheating and conserved much-needed water. In other words, the sun's rays, instead of falling on their backs, would have fallen vertically on our ancestors' heads – a far smaller area. Wheeler's research – using models of *Australopithecus afarensis* – suggests that walking on two feet could have reduced heating by as much as 60 percent. In addition, in an upright stance far more of an animal's body is raised above the hot ground and away from the heat radiating from it, while further cooling would have been achieved through contact with breezes and air currents found several feet above ground level. "By walking on two feet, humans developed the animal world's most powerful cooling system," says Wheeler.[25]

These are all intriguing suggestions, although there is a distinct lack of hard evidence to support any of them, for this period of our prehistory is meagerly represented in terms of fossils. Only two species are known to predate *afarensis*. The first – *Ardipithecus ramidus* – is thought to be about 4.4 million years old. Its remains consist of teeth, bits of skull, arm bones, and a partial skeleton found in Ethiopia by a team led by Tim White of the University of California, Berkeley. These fossils are now being analyzed by a multidisciplinary team of international scientists, including White, and should shed important light on this dim recess of our evolution. To judge from early reports, *ramidus* is more like an ape than a hominid, although its foramen magnum – the hole in the base of the skull – is situated fairly far forward, an indication of bipedalism. However, we will have to wait for further studies to be published to determine whether it was a direct ancestor of australopithecines, and therefore of ourselves, or just an offshoot from the "bush of life" that was destined for early extinction. White believes *ramidus* is a human ancestor, indeed the earliest-known member of our lineage, but until further analysis and excavations are completed, researchers cannot be sure. In short, science has yet to reach its verdict on *ramidus*.[26]

Discovery of the other predecessor of *afarensis* is the handiwork of Meave Leakey – head of paleontology at the National Museums of Kenya, wife of Richard, and daughter-in-law of Mary. She recently found fossils that are attributed to an *afarensis* ancestor – and which have been dated as being at least 4.2 million years old – at a site called Kanapoi, near Lake Turkana in northern Kenya. Lake Turkana, which is fed from the north by the Omo River, lies at the very heart of the Rift Valley, and it is clear that in the surrounding savanna, many of the great sagas of human evolution were resolved, as we shall see in future chapters. "Almost certainly our first apelike ancestors emerged in Africa, and few places offer as rich a fossil record as this region," says Meave Leakey. "Tectonic activity has uplifted ancient sediments, exposing to rapid erosion the soils in which the early hominids' bones were fossilized. Thus each rainstorm can bring new fossils to light. In addition, volcanism over the eons has deposited many layers of ash. Radioactive minerals in the ash decay at known rates, letting us date each layer and the fossils in between."[27]

Kanapoi is a parched and barren landscape riddled with deeply eroded gullies, and in one of these Meave Leakey and her colleagues discovered pieces of the jaw, lower face, and leg bone of a hominid that had some apelike features, and some that were definitely humanlike. The mandible (lower jaw) is similar to a chimp's, for example – a tight u-shape, while a human's widens at the back of the mouth. However, the leg bone is strong and its knee and ankle joints are similar to those of *afarensis*, indicating

The searing desert of Kanapoi where Meave Leakey uncovered the first evidence of 4.2-million-year-old *Australopithecus anamensis*. The sediments of Kanapoi are a goldmine for paleontologists, erosion and tectonic activity having exposed the ancient layers of soil and rock in which fossils had lain undisturbed for millions of years.

that, even 4.2 million years ago, our ancestors were well equipped for an upright life. She named this new creature *Australopithecus anamensis* (*anam* is the local word for lake), and it remains the earliest-known member of our lineage, the immediate ancestor of the *afarensis* hominids who left their footprints at Laetoli. Subsequent research at the nearby site of Allia Bay has also uncovered *anamensis* fossils, although it seems the species was not living only in open country. Seeds and bones of forest monkeys and antelopes have been found in the same sediments, suggesting that this early progenitor of humanity was still inhabiting closed-canopy woodland some of the time.[28]

Indeed, its successor, *afarensis* – which probably evolved from *anamensis* a few hundred thousand years later – also seems to have been a woodland creature. Animal and plant fossils at the various hominid sites now point to environments that include open woodland, gallery forest, closed woodland, and open grassland. Our predecessors did not simply stride out onto open savanna. It was a fairly gradual transition. "Life would have been dangerous away from the forest for small-bodied and relatively slow-moving creatures such as these, and it would be surprising if they had not regularly sought the shelter of their former home," says Ian Tattersall.[29] While its feet, legs, and hips were attuned to upright walking, *afarensis*'s arms were long and powerful, rather like an ape, suggesting that it still climbed trees fairly regularly. "These early australopithecines had bodies that were a mosaic of human and apelike characteristics," says Richard Potts, director of the Smithsonian's Human Origins Program, in Washington. "They were adapted both to walking on two feet and to tree-climbing." And the

Meave Leakey with colleagues (top) in 1993, studying some of the samples of *anamensis* that were discovered at Kanapoi (center). Her finds included a complete lower jaw (below right) that closely resembles that of a modern chimpanzee. However, the fragments of tibia found (bottom left) are those of a creature that evidently walked on two legs.

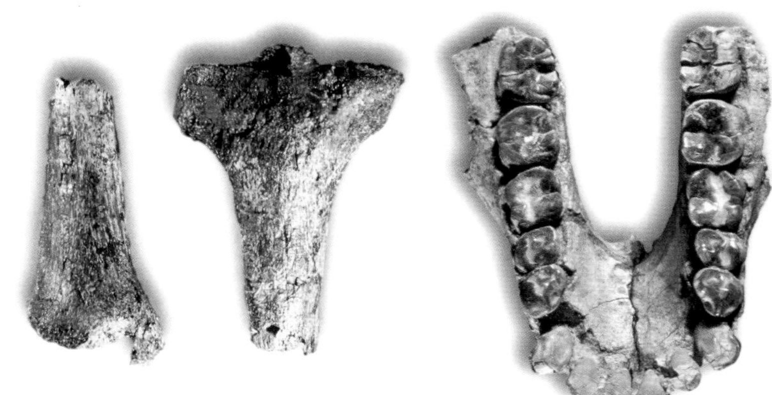

BONES OF CONTENTION

Fossil hunters have few rivals when it comes to scholarly sniping, as *Newsweek* once remarked. More than any other scientific profession, paleontologists have a poor reputation for squabbling. And it is true that many of the distinguished names which feature in this book have, at one time, indulged in a little acrimonious bickering. These disputes range from the disagreements of Raymond Dart and Sir Arthur Keith over the provenance of the Taung child (see page 41), to arguments that took place between Richard Leakey and Donald Johanson over the definition of *Australopithecus afarensis*, and even to current altercations (which we shall encounter later in the book) about the origin of modern human beings. These disputes have occasionally become rather heated and it would be easy to conclude from them that the profession is made up of some unusually hot-headed, inflexible individuals.

In fact, the behavior of paleontologists is no worse than that of any other group of scientists, despite appearances. Part of the trouble lies with the very nature of the subject that they study – human beings. Although poorly funded compared with other sciences, paleontology attracts great media interest, and interpretations of fossil finds often make headline news. Arguments are often simplified and distorted in the process, and protagonists find themselves being pressured into holding increasingly polarized views.

In addition, there is a distinct paucity of evidence from which to make arguments. Our understanding of our evolution is based on the discovery of the remains of only a couple of thousand cavemen or apemen, most of whom have left a mere scrap or two of bone to posterity. In other words, there is only a limited amount of data – but a correspondingly large opportunity for speculating.

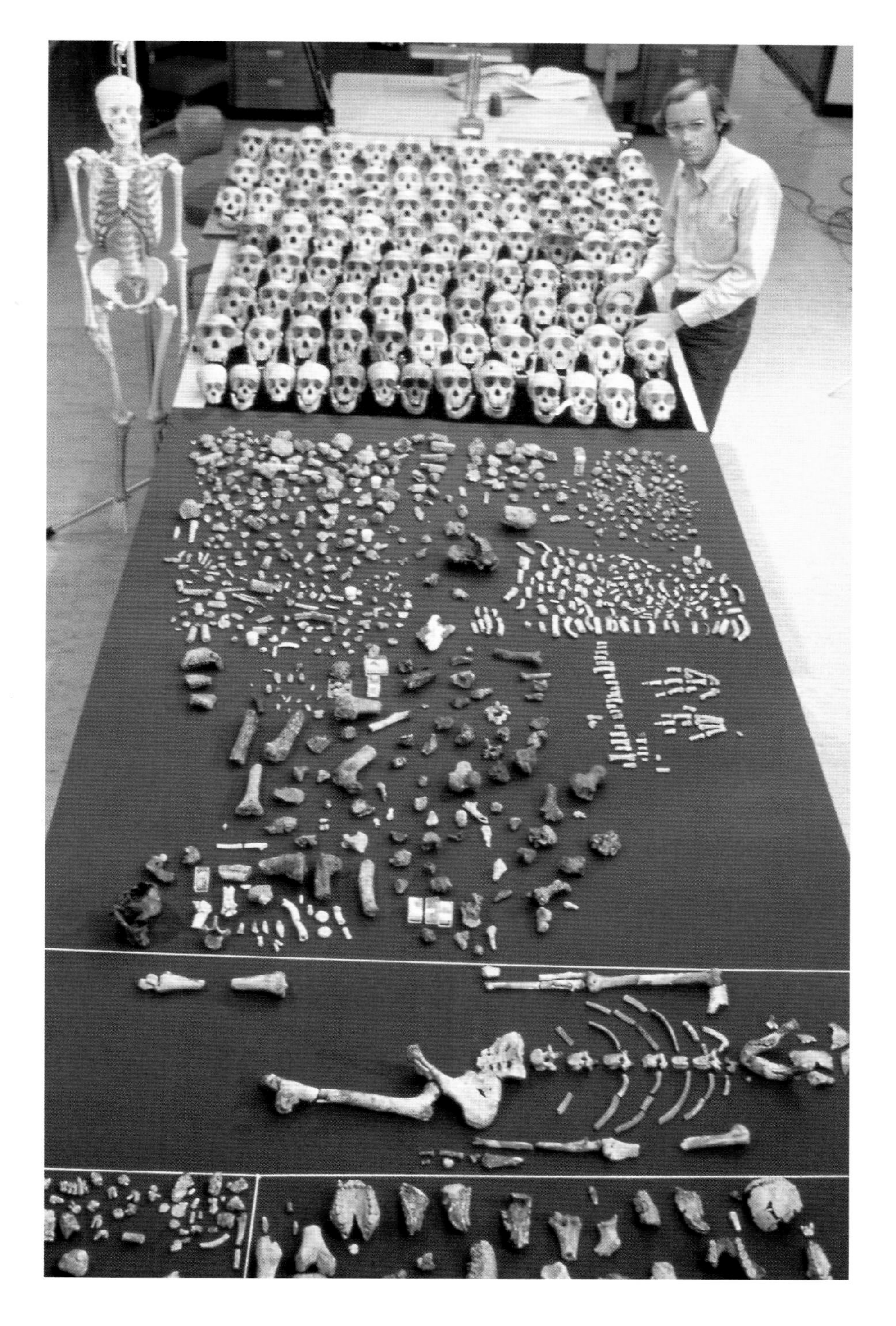

OPPOSITE, FAR LEFT: Louis Leakey who startled many fellow scientists with his claim that *Homo habilis* was the first member of the true lineage of humans.

OPPOSITE LEFT: Richard Leakey, who initially doubted the idea that *Australopithecus afarensis* was a true separate species as Don Johanson and Tim White suggested.

LEFT: Tim White with the complete Hadar and Laetoli fossil collections, including Lucy and the First Family, which he and Johanson used as the basis for naming *afarensis* as a new species.

BELOW: Don Johanson subsequently fell out with Tim White, and now directs the Institute of Human Origins, at Arizona University.

reason they possessed this pick-and-mix of ape and human attributes had much to do with their environment. "Australopithecines needed locomotor versatility. They had to live both in densely wooded areas and quite open habitats," says Potts.[30] We can see some evidence for this open-ground perambulation at Laetoli, as Andrew Hill points out. "Most other hominid remains that have been found tend to be associated with streams and lakes, and have been preserved by their sediments," he says. "But at Laetoli there are no prints of hippos or crocodiles, for one thing, and no sign of lake sediments. This suggests these people may have been walking around on open ground a fair bit."[31] However, such was the climatic variability of the time that our apemen ancestors could never be sure of continued tree protection. Variable behavior was their response, as we can see in their pelvises, which were adapted to upright gait, and in their long, powerful arms, which reveal a continued reliance on life in trees for at least part of their lives.

This arboreal tendency would have been particularly strong in a species that was clearly still a herbivore. "*Afarensis* evolved out of a vegetarian past, and although they may have occasionally eaten termites, lizards, or other small creatures, they relied primarily on vegetation," says Johanson. Australopithecines were certainly not the masters of their environments, and were still more likely to be preyed on than to prey. If anything gave them an edge over other animals, it was living in social groups of up to 30 individuals, Johanson believes. "The group would have offered common defense against carnivores, especially at night without protection of fire. I can

OPPOSITE: Sunrise at Hadar in Ethiopia where Donald Johanson and his team made some of paleontology's most spectacular and important discoveries about the early origins of our ancestors.

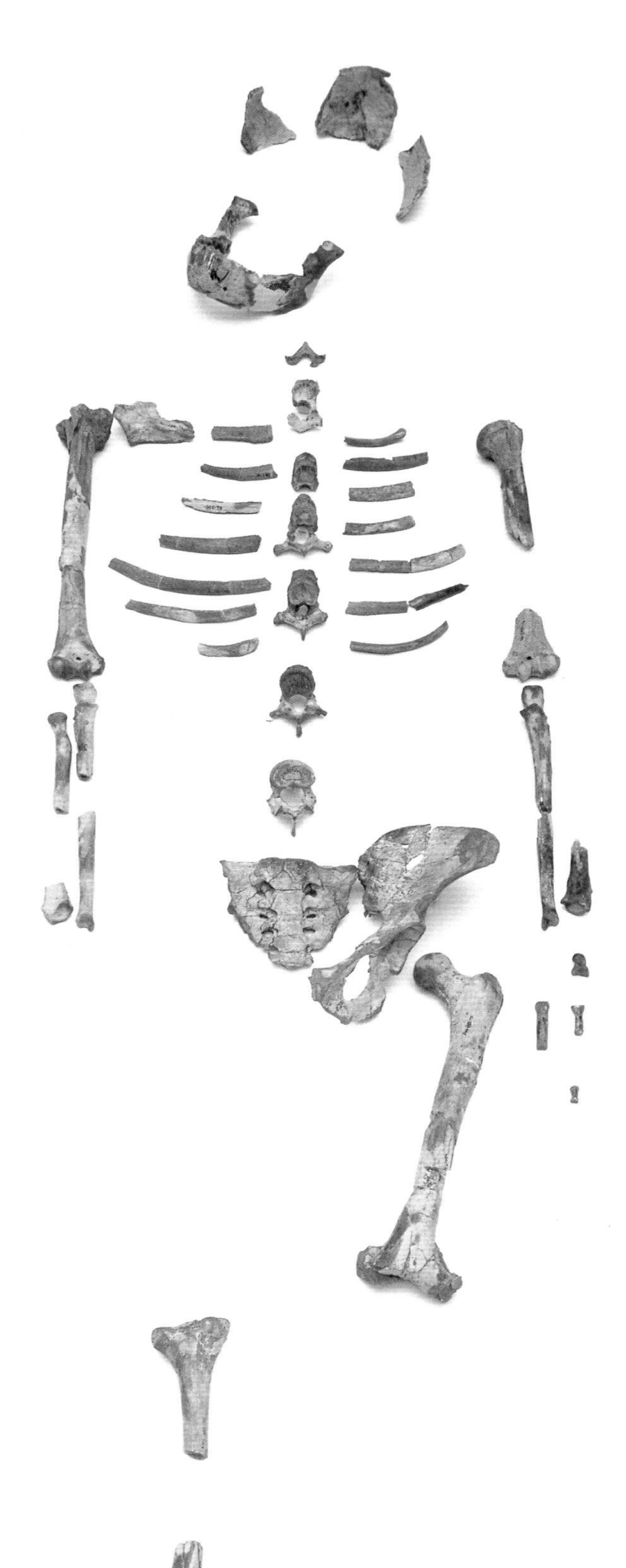

'Lucy': her remains comprise a 40-percent complete skeleton and include skull fragments, a mandible, and most of the left and right arms. The skeleton is assumed to be that of a female because of its small size.

imagine a group of howling *afarensis* parents throwing rocks at a threatening saber-toothed cat. Carnivores could get more meat for much less trouble by running down a gazelle."[32]

Clearly, *Australopithecus afarensis* was no pushover. Indeed, it can be regarded as one of our planet's great primate survivors, having thrived for a million years after making its first appearance on the fossil scene about 4 million years ago. It would certainly be wrong to dismiss it as some kind of halfway hominid house on the route to humanity. It was a well-established species in its own right, and showed great adaptability, which undoubtedly contributed to its endurance, as Johanson emphasizes. And he should know, for he was responsible for discovering the most complete skeleton of this ancient human forebear: Lucy. She was to become the first true celebrity among fossils. Indeed, as Johanson ruefully acknowledges, Lucy is better known than her discoverer, and has appeared in cartoons, crossword puzzles, rock lyrics, plays, and even as a tattoo. Her 3.2-million-year-old skeleton was discovered on November 24, 1974 in a maze of ravines near the River Awash in the Hadar region of Ethiopia's Afar Triangle. And, of course, luck played a part. Johanson had intended to spend the day bringing his field notes up to date but was persuaded by one of his students, Tom Gray, to look for fossil animal bones among the rolling hills of scorching sand, silt, and ash of Hadar. As they walked through this blistered terrain, the pair saw a fragment of arm bone that clearly belonged to an apelike creature. "We looked up the slope," recalls Johanson. "There, incredibly, lay a multitude of bone fragments – a nearly complete lower jaw, a thighbone, arm bones, ribs, vertebrae, and more! Tom and I yelled, hugged each other and danced, mad as any Englishmen in the midday sun."[33]

Since his discovery of Lucy in 1974, paleontologist Don Johanson continued to excavate at Hadar, where he made many more important discoveries, including the 'First Family', the remains of 13 *Australopithecus afarensis* individuals found buried together.

Johanson and Gray roared back to camp in their Land Rover, horn blazing, delirious with their discovery. Beer was cooled in the Awash River and a special goat barbecue was laid on to celebrate. In the end, a total of 47 bones belonging to one individual were dug up – nearly 40 percent of a single skeleton. Her pelvis shape indicated she was female, while her short legs suggested she had been only 3.5–4 ft (1.1–1.2 m) tall. She had already cut her wisdom teeth, which indicated that she was an adult when she died. In addition, the lipped edges on several of her bones revealed that she had suffered from arthritis. "Surely such a noble little fossil lady deserved a name," recalls Johanson. "As we sat around one evening listening to

Beatles' songs, someone said: "Why don't we call her Lucy? You know, after 'Lucy in the Sky with Diamonds.' So she became Lucy."[34]

Lucy was a paleontological godsend, and her skeleton has spawned a quarter century of vigorous scientific debate. For the next four years Johanson continued to hunt for fossils in the Afar region with remarkable success, uncovering a series of Lucy-like creatures. The most important of these finds became known as the First Family, and was discovered in 1975 by one of Johanson's students, Michael Bush, at a site simply labeled Afar Locality 333. Bush spotted an upper-jaw fragment, with teeth attached, sticking out of the ground. Then a thigh and heel bone were spotted, followed by more leg bones. More and more fragments came to light. 'The entire hillside was dotted with bones,' says Johanson. In the end, the remains of 13 individuals – nine adults and four children – were discovered at site 333, a count based on the number of different teeth and pieces of jaw found there.[35]

Apart from the sheer number of fossils (more than 200 hominid fragments were recovered), these remains possessed several intriguing features. There were no gnaw marks made by carnivores on the bones and no signs that they had been weathered before being buried, suggesting the group had been entombed very rapidly – probably by some sudden catastrophe. The most likely candidate is a flash flood, Johanson believes. And if this group of adults and children had been together at the same place and same time before being struck down, they were very likely all members of the same extended family. So he called them the First Family.[36]

It was a discovery of crucial importance. Scientists had been puzzled by the fact that all the Lucy-like fossils that they had dug up in Ethiopia, Kenya, and Tanzania over the previous few years were of very variable sizes. Lucy was very small, coming in at under 4 ft (1.2 m) tall, while other fossils suggested their owners were more than 5 ft (1.5 m) tall. Was this a single stock, but one that varied considerably, or were scientists looking at two different species? The First Family provided crucial clues because their bones showed enormous differences in physique among the adults, suggesting that scientists were dealing with a single species, one that displayed dimorphism – a phenomenon typified by large males and small females. In this case, the bodies of males must have averaged about 4 ft 10 in (1.5 m) tall and 143 lb (65 kg) in weight, while females were on average only about 3 ft 3 in (1 m) tall and 66 lb (30 kg) in weight. And that dimensional diversity in turn suggests a very nonhuman, nonmonogamous social structure. Just like male gorillas, men of the First Family may

OPPOSITE: A recreation of a female *Australopithecus afarensis*. "Lucy" may have looked something like this.

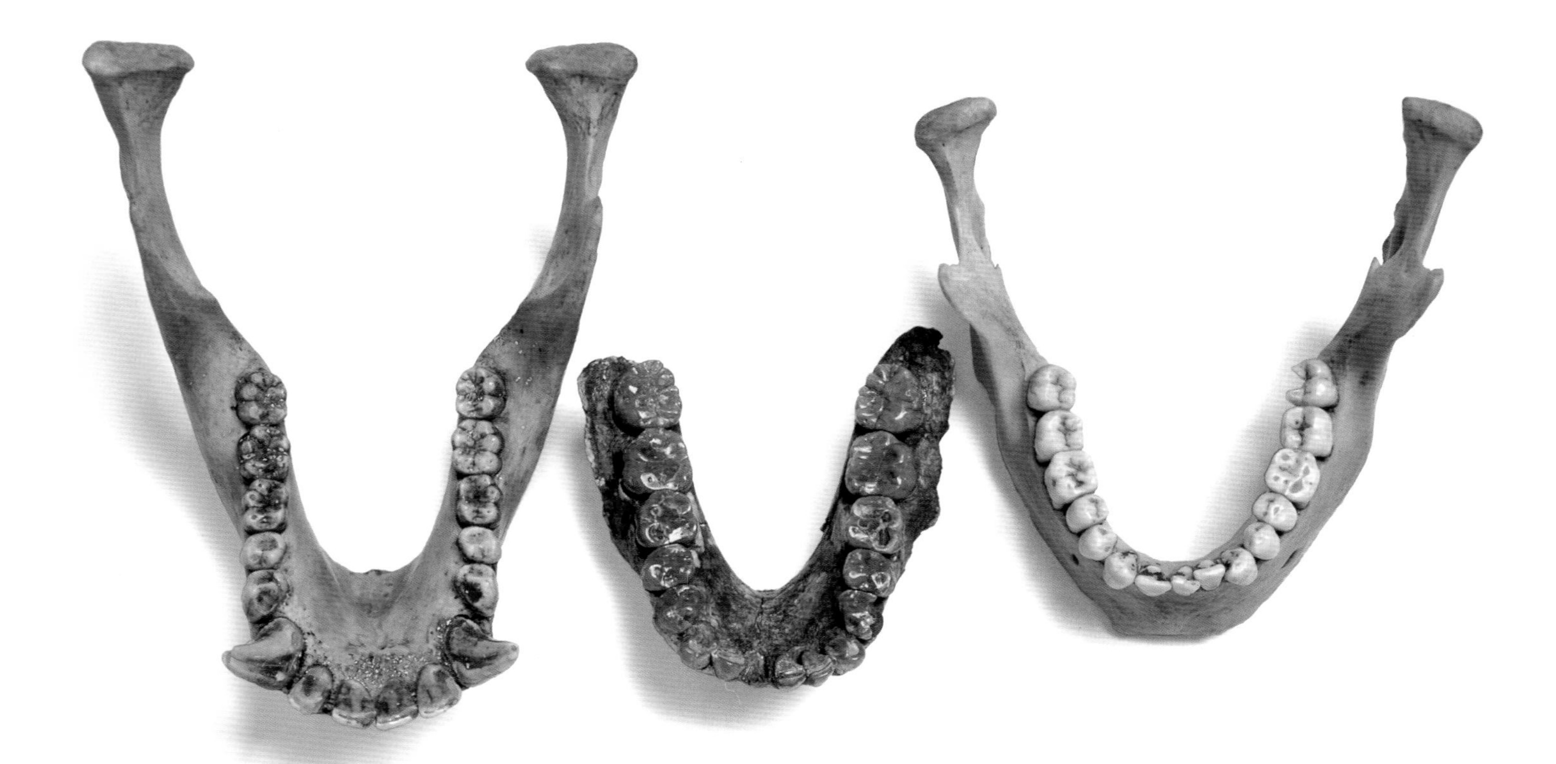

The lower mandible of a modern human (right), *Australopithecus afarensis* (center) and a chimpanzee (left). The australopithecine jaw is an intermediate shape between the parabola of *Homo sapiens* and the tight u-shape of a chimpanzee.

have developed large bodies to compete with each other for access to large groups of females (harems). However, male gorillas have large, powerful canine teeth to fight off rival males, and it is significant, says Johanson, that *afarensis* canines are much smaller. Perhaps the species was slowly evolving away from an apelike lifestyle and beginning to adopt a more monogamous mating system.[37]

It was on the basis of these ideas that Johanson, together with Tim White, announced at a 1978 Nobel symposium in Stockholm that Lucy, the people of Laetoli, the First Family and several other fossils belonged to a single species: *Australopithecus afarensis*, the Southern Ape from Afar. It was a controversial claim, for not every scientist was convinced by Johanson's deductions. However, most paleontologists now accept that his classification was correct and that he had at least simplified the picture of our early evolution.[38]

Both Lucy and the First Family have been dated as being about 3.2 million years old, and their species, says Johanson, was probably the common ancestor of all subsequent branches of humanity, including our own. Lucy, in a sense, could have been the mother of humankind.[39] But by about 3 million years ago, *afarensis* disappears from the fossil record. In its place, new apemen would appear – but these would not be so easily categorized in terms of a single species. Variations on the *afarensis* theme were about to be played out across much of Africa and in many different ecological niches. A multiplicity of human predecessors would make their appearance, strut and fret their hour upon the stage of evolution, and then be heard no more. This flowering of different hominid forms is what we should expect, as Cambridge University anthropologist Robert Foley stresses. "Evolution is about diversity. It's not just lines leading to us."[40] This is an important theme that we shall explore in detail at the beginning of the next chapter, before going on to discuss the fate of *afarensis*, the species that followed in its wake, and the highly gifted and decidedly unusual human beings who have played such an important part in unraveling this intriguing and mysterious part of our history.

CHAPTER TWO

THE STAR WARS BAR

The past century has seen an astonishing growth in our understanding of our evolution. We have tracked down primeval pieces of our predecessors and used them to create a 5-million-year-old family tree for humanity. It is an extraordinary achievement, given that it is based on the discovery of little more than 2000 individuals – for that is the tally of all the apemen and human beings whose prehistoric bones have ever been dug up and studied over the past 100 years.[1] Since there are now more than 6 billion people living on our planet, and we have been evolving as humanlike creatures for so long, such an assembly of fragments is modest to say the least. Nevertheless, through scrupulous analysis, scientists have pieced together a convincing story, aided by molecular biological analysis of ourselves and fellow primates, and by examining our antecedents' tools and weapons.

Bones, stones, and genes: it is a powerful combination, although the first on this list still holds primacy as a source of knowledge. Fossils are the most solid manifestations of our lost ancestors. As Don Johanson describes it, "Seeing a complete skull is like looking at a person."[2] Understanding exactly what a skull or a bone or a tooth can tell us is not such an easy matter, however. Considerable judgement is needed to make sense of fossils and to put them in an intelligible framework. Unless that is done, a fossil is just an old bone. Unfortunately, it is this interpretation that can often divide paleontologists. For example, some believe the human lineage is extremely complicated, filled with a multiplicity of different hominid species. These scientists are known as "splitters" because they like to bisect the pathways of human evolution as often as possible and tend to be free in their creation of new species. At the other end of the spectrum are the "lumpers." They like to keep our genealogy simple and straightforward, and are inclined to categorize different fossils under one species name.

A classic depiction of the evolution of mankind as a simple march of progress, a convenient but erroneous view. In reality the Africa of some 2 million years ago would have been home to several different hominid species, some of which turned out to be dead-end branches on the bush of human evolution.

Lumpers and splitters: it is a basic division within the science, although – occasionally – it is possible to be both. Consider the example outlined in the previous chapter – of the hominid bones that Johanson found in Hadar, and those that Mary Leakey discovered at Laetoli. Paleontologists looked at these remains and saw similarities. However, they also noted that the fossils came in various shapes and sizes. So was this one or two species? Then Johanson made his announcement with Tim White in 1978 that these remains all represented just one type of humanity: *Australopithecus afarensis*. In ascribing these finds to a single species, Johanson was, in one sense, being a lumper. He was simplifying our genealogy. However, he was also claiming that *afarensis* was a new species, never previously classified by science. He was therefore stressing the innate difference between its members and previously identified hominids. So, in that sense, he was adding a new branch to the tree of evolution: he was being a splitter.

For the first half of the 20th century, the splitters' camp held sway; virtually every time a fossil was unearthed, it was assigned to a new species, a practice that rapidly got out of hand. There seemed to be more species than paleontologists to investigate them: *Australopithecus transvaalensis*, *Paranthropus crassidens*, *Homo kanamensis*, and many more. By the middle of the century, the picture had become confused and lumping became popular, eventually reducing the number of hominid species to a handful. This had the merit of making our evolution easier to understand, but, taken too far, had one serious drawback. By coalescing so many different species, by slashing off so many branches from the bush of human evolution, the route from *afarensis* to *Homo sapiens* looked more or less like a straight line, not a complex tree.

Such a simplistic vision only enhanced an existing, mistaken belief that our emergence was a predestined outcome of primates leaving their trees. Without any evidence of deviation from the straight and narrow of natural selection, our evolutionary tree was reduced to a trunk, and the arrival of *Homo sapiens* looked preordained. Our early ancestors were made to look like biological heroes impelled toward betterment – in contrast to other primates, who lolled about in the trees, "satisfied with their circumstances," as the anatomist and anthropologist Sir George Elliot Smith put it in the early 20th century.[3] Or, as his US counterpart Ray Chapman wrote, "Hurry has always been the tempo of human evolution. Hurry to get out of the primordial ape stage, to change body, brains, hands, and feet faster than it had ever been done in the history of creation. Hurry on to the time, when man could conquer the land and the sea and the air; when he could stand as Lord of all the Earth."[4] This perception of purpose and striving in human evolution was a common attitude and still is, as Stephen Jay Gould emphasizes. "The march of progress is the canonical representation of evolution – the one picture immediately grasped and viscerally understood by all." Indeed, the idea of linear advance has become so entrenched that it has become – at least until relatively recently – a mental straitjacket for our understanding of evolution. "The word itself becomes a synonym for progress," he says. But life is not like this. It is "a copiously branching bush, continually pruned by the grim reaper of extinction, not a ladder of predictable progress", as Gould says.[5]

Certainly, there was nothing predestined in the appearance of *Homo sapiens* (as we shall see later), although it is perhaps natural that we assign exclusive, preordained greatness to our own species, albeit mistakenly. And that is especially true given the absence of other hominids on our planet. We take it for granted that we are the only ones. We feel special. Yet it is probably for only the past 30,000 years – a tiny fraction of our evolutionary history – that Earth has been home to only one species of

Makapansgat cave, one of the system of limestone caverns in South Africa in which apeman remains (such as those of the Taung child), more than 2 million years old, have been uncovered.

hominid. "We have got to get rid of an idea, now deeply ingrained in our conscious, that because there is only one species of human being today, this has always been the case," says the Israeli paleontologist Yoel Rak. "For most of our evolution the opposite was true. Think of that scene from *Star Wars* – in the bar where you see all kinds of aliens playing and drinking and talking together. I believe that image gives a better flavor of our evolutionary past."[6] This point is backed by Tattersall. "It should never be forgotten that the history of the human family eloquently demonstrates that there are many ways to be a hominid and that our way is but one of them."[7] Or, as Gould stresses, "The current status of humanity as a single species, maximally spread over an entire planet, is distinctly odd."[8]

In fact, our prehistory was one in which the landscapes of Africa, and later the Old World, were filled with different hominid species thriving – or at least surviving – at the same time. This vision is an important one to keep in mind, particularly when we come to examine *afarensis*'s disappearance from the fossil record about 3 million years ago. The question is: what came next? A great deal is the answer, although the story of the subsequent evolution of our lineage is less straightforward, and certainly more geographically diverse, than its opening stages (as far as one can tell from the current fossil record, that is). Indeed, the next part of our saga opens not in northern East Africa, but at the other end of the continent, in South Africa.

Raymond Dart with the Taung child skull. Despite the derision of the scientific community, his claim that the skull was that of an ancestor of modern humans was subsequently vindicated.

In 1924, Raymond Dart, the newly appointed Professor of Anatomy at Witwatersrand University, Johannesburg, discovered the front half of a child's skull, with jaws and teeth, of an odd-looking ape. The fossil had apparently rested on the desk of the manager of a lime works at Taung, near Kimberley, until one of Dart's colleagues spotted it and sent it to Dart, along with other remains from the site. Dart later recalled receiving the box as he was dressing, in white tie and tails, for a wedding. With his collar undone, Dart opened the box and pulled out the Taung skull. "I stood in the shade holding the brain as greedily as any miser hugs his gold," he recalled. "Here, I was certain, was one of the most significant finds ever made in the history of anthropology. These pleasant dreams were interrupted by the bridegroom himself tugging at my sleeve."[9]

The Taung child: the find consisted of pieces of skull, a jawbone and an endocast – petrified sediments that had filled its cranium. Although superficially apelike, Dart realized that the skull also had humanlike features.

Dart, an ambitious and highly gifted Australian anatomist who had a fascination with fossils, had spotted several key features in the skull. However, Johannesburg was scarcely more than a frontier town in those days, and Dart had no sophisticated instruments to use for his investigation, so he borrowed a set of his wife's knitting needles and slowly began to pick away at the encrusted skull. He found that the teeth were unlike those of any ape, while the foramen magnum (the hole through which the spinal cord passes) was located in the center of the base of the skull – a feature of an animal that walks on two legs. In addition, a cast of the inside of the Taung skull revealed that a gap in the rear of the child's brain (the lunate sulcus) was in a different position from that of an ape and was situated more like that of a human brain. Dart called his new discovery *Australopithecus africanus* (southern ape of Africa) – the first time the name *Australopithecus* had been applied to early hominids.[10] Although he recognized that adult features had yet to develop on the skull (which, he said, came from a six-year-old to judge from the fact that the first permanent molar teeth had just erupted), he was not restrained in claiming that the species was a predecessor of modern humans, was intelligent, and made tools.[11] The discovery made headlines around the world but was treated with scepticism by the scientific establishment. The distinguished and highly influential anthropologist Sir Arthur Keith was particularly scornful. The Taung skull had belonged to a young ape "showing so many points of affinity with ... the gorilla and the chimpanzee that there cannot be a moment's hesitation in placing the fossil in this living group", he wrote in *Nature*. As for its claim to be a forebear of the human race, that was like claiming "a modern Sussex peasant as the ancestor of William the Conqueror."[12] At best, African australopithecines were only apelike remnants, left behind as human evolution unfolded in Asia, it was said.

Now this idea that the crucible of human evolution was Asian, and not African, may seem odd given that Charles Darwin had stated in *The Descent of Man* that "it is somewhat more probable that our early progenitors lived on the African continent than elsewhere,"[13] and that the work of Johanson, the Leakeys, and many other paleontologists have since demonstrated through their uncovering of abundant fossil evidence that we are, above all, a species of African origins. But Africa was out of fashion in Dart's day, and scientists were fixated by the idea that the home of mankind was Asia. It was the biggest continent, had the richest variety of life forms, possessed the oldest civilizations, and was blessed with the right climatic conditions to enforce our descent from trees so that we could adapt, and "progress," to our current status, it was argued.

With the benefit of 20:20 hindsight, we now know that Dart was right in many ways and that his scientific peers were wrong. Africa is now recognized as the principal stage on which the drama of our early evolution was acted out, and we now believe *africanus* was our ancestor; its remains are certainly much more ancient than the human fossils of Europe and Asia. On the other hand, this species could scarcely be described as being particularly intelligent, nor has any evidence ever been uncovered that it made tools. Indeed, *africanus* was in many ways like *afarensis*, displaying a similar body and brain size (although the differences between males and females seem less marked). The front teeth were a little smaller, the back teeth relatively larger; the face a little flatter, and the cheekbones more prominent. And, like *afarensis*, *africanus* walked upright and probably foraged for a mixed but mainly vegetarian diet, although there may have been some scavenging for meat.

For years, Dart's discovery languished in intellectual limbo, a victim of the derision of scientific orthodoxy. Indeed, it would have remained there for much longer had it not been for the efforts of Robert Broom, a remarkable Scottish doctor then working in South Africa. Broom became one of the few scientists to support Dart. A graduate of Glasgow University, he was described by the great UK geneticist J.B.S. Haldane as a man of genius, fit to stand beside Shaw, Beethoven, and Titian, while his biographer, G. Findlay, observed that he was about as honest as a good poker player – an intriguing mix of characteristics if nothing else. Broom was also obsessed with the benefits of sunlight, with the result that he used to remove his clothes while hunting fossils in remote hot places to maximize his exposure to the sun's rays, unfortunately misplacing his attire completely on one occasion.[14] Although he practiced medicine, Broom also ran a thriving wholesale fossil-export business, hawking valuable remains

to American museums, which somewhat disgruntled the South African government, who accused him of selling off national assets. For a while he was held in bad odor by the local authorities. More importantly, however, Broom was an enthusiastic supporter of Dart, or, to be more precise, of what Dart had discovered. Two weeks after the announcement of the discovery of *africanus*, he barged into Dart's laboratory unannounced, strode past both professor and students, and knelt in silent adoration before the Taung child's skull.[15]

Broom also proved to be an exceptionally good paleontologist and, in 1938 at the age of 70, he found the first of a new type of australopithecine, before going on to find several more important fossils of similar-looking hominids. An enthusiastic splitter, he gave them all different names. Today, they are simply known as a single species: *Australopithecus robustus*, a brawny apeman that was slightly larger than *africanus* and had molars that acted like a battery of flat grinders. Think of *robustus* as a heavy-duty masticator of roots, shoots, and other food. Broom also found more remains of

Robert Broom (below, left), one of the few scientists who believed that Raymond Dart's original claims about the Taung child were correct. His subsequent discoveries, including those of *Australopithecus robustus* (below), and his investigations at sites such as Sterkfontein (opposite), not only showed that Dart was indeed correct but also irrevocably changed our understanding of our ancestors.

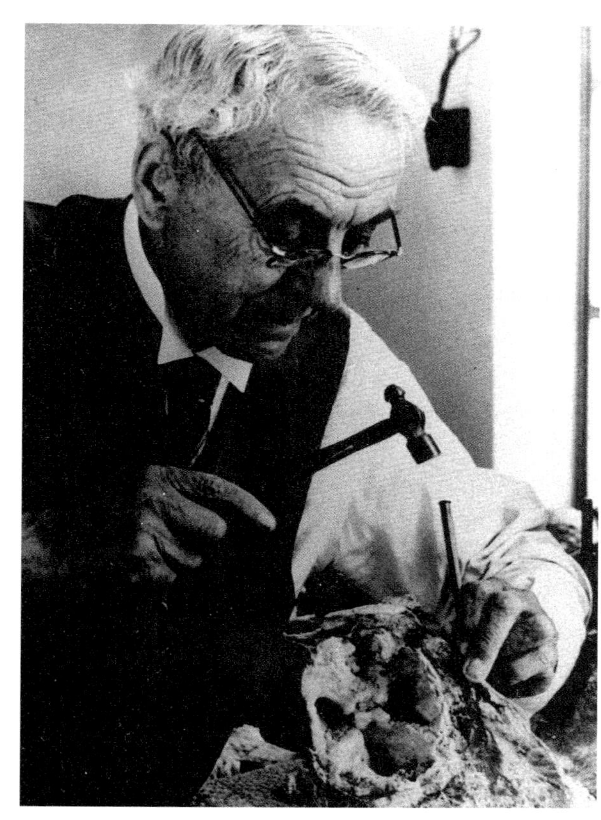

africanus and, in 1946, published a massive monograph on the australopithecines. In it he stated "these primates agreed closely with man in many characters. They were almost certainly bipedal."[16] Broom kept excavating and, in 1947, discovered the skull, pelvis, and spine of an australopithecine that clearly walked upright. The scientific establishment capitulated – 23 years after Dart's original discovery. "Professor Dart was right, and I was wrong," Keith admitted in a letter to *Nature*. Dart may have been correct in the first place, but it was Broom who proved the truth of his cause. By now Broom was 84 and anxious to complete his final monograph on the species. On April 6, 1951, after making final corrections to his manuscript, he announced: "Now that's finished … and so am I." He died that evening.[17] "More than any other scientist or discovery before that, Broom's work on *Australopithecus* fundamentally and irrevocably revised the study of fossil man," says John Reader in his history of human paleontology, *Missing Links*.[18]

In fact, Broom had made several critical achievements. He had established that australopithecines were credible ancestors of humanity, even though they were small-brained. Bipedalism, not intellect, was the first defining characteristic of our lineage, it was now becoming clear. Broom had also provided convincing evidence that Africa, not Asia, was the setting for the early evolution of *Homo sapiens*. In addition – and most importantly for our continuing story – he had found the first really convincing evidence of the multiplicity of hominid species. It is now clear that about 2 million years ago there were at least two types of hominid that had succeeded *afarensis* and were thriving in different ecological niches: *Australopithecus robustus*, which probably lived by devouring high-bulk, low-nutrition food, including leaves and shoots, and *Australopithecus africanus*, which existed on more varied fare, possibly including meat. The *Star Wars* bar of mankind had started to fill up.

However, *africanus* and *robustus* were not to be the only hominid saloon visitors to turn up, it transpired. In 1959, the industrious Mary Leakey and her husband uncovered in Kenya the fossilized remains of a third australopithecine, now known as *Australopithecus boisei*. This also appears, like *robustus*, to have been a heavily built vegetarian, possessing enormously thick jaws and large back teeth.[19] In other words, now we were three: *africanus*, *robustus,* and *boisei*, although the last two seem to have been virtual mirror images of each other, both having adopted the same ecological niche. And a good one at that, for both survived from about 2.5 to 1 million years before present – the former in South Africa, the latter in East Africa – although their lineage did eventually peter out. Their brains were roughly the same size as that of

Australopithecus africanus, despite both having larger faces, jaws, and teeth.

From all this evidence, paleontologists have drawn up a fairly standard picture of early human evolution that leads us from the hominids of the last chapter to the apemen of this one. Firstly, *Australopithecus anamensis* evolved into *afarensis*. *Afarensis* not only led to *Australopithecus africanus*, it also branched off to produce *robustus* and *boisei*. These variations of the australopithecine theme then flourished in Africa until *robustus* and *boisei* disappeared from the fossil record about a million years ago, giving their like a highly creditable survival rating. Upright posture and small brains seem to have been the leitmotifs for most of our evolutionary history, in fact. And as for *africanus*, like its *afarensis* predecessors, it appears to have been mainly a creature of forest fringes, woodlands, and open grassland, although evidence from one site (Makapansgat in South Africa) suggests it may sometimes have lived in dense forests. *Africanus* not only obtained fruit and leaves from trees but probably also dug up tubers and roots, and may even have begun scavenging meat from dead animals, its behavior being a more flexible response to the changing environments at the time.

This picture, which is accepted in varying forms by most scientists, begs a key question: what happened to *Australopithecus africanus*? *Robustus* and *boisei* were dead ends. But what happened to Dart's southern apeman? Fossil records disappear around 2.4 million years ago. Where did *africanus* go? It took the involvement of another iconoclast to unearth the first convincing answers to these questions – Louis Leakey, one of the greatest of all fossil-hunting adventurers.

TOP: Mary Leakey excavating at the Olduvai Gorge in the company of the family's two faithful dalmations. ABOVE: Australopithecines were forced to broaden their diets in response to the changing nature of their traditional habitats. At this point, in addition to their largely herbivorous diet, they may have started scavenging meat from dead animals.

APEMEN: MURDERERS OR VICTIMS?

Raymond Dart, discoverer of the *Australopithecus africanus* Taung child skull in 1924, was spectacularly correct when he said our origins were African. Some of his other speculations were very wide of the mark, however, and in one case triggered dangerous misconceptions about human behavior. Dart believed that *Australopithecus africanus* made weapons and used them to kill animals – and also to murder, and eat, their rivals. They were "carnivore creatures that seized living quarries by violence, battered them to death, tore apart their broken bodies … and greedily devoured their writhing flesh," he wrote. Dart based his arguments on damaged skulls and bones found at Taung and other sites. Holes in braincases – like those in the Taung child – were the result of other australopithecines bashing their fellow apemen with stone tools, he said.

Dart's ideas were taken up in the 1960s by the US writer Robert Ardrey in his book *African Genesis*, which claims that mankind developed big brains only after it had begun making implements for killing. Our predecessors then needed increasingly sophisticated neurologic controls in order to handle increasingly complex weapons. The idea that the urge to kill drove our intellectual development was further promoted by Stanley Kubrick and Arthur C. Clarke in *2001:*

A Space Odyssey, in which an apeman is seen playing with bones and suddenly realizes he can use one as a weapon. A femur is sent spiraling into the air and is transformed into a spinning spaceship. Murder led to the development of technology, in other words.

This premise is based on false interpretations of fossil finds, however. The holes in the *africanus* skulls were not caused by their neighbors bashing them. They were inflicted by animals. Carnivores, such as leopards, who dragged off dead apemen, left this grim stigmata. In addition, the Taung child is now thought to have been the victim of an eagle that carried off its body, or at least its skull. *Africanus* was not the hunter. It was the prey. Murder is not in our genes.

Double indentations on an australopithecine skull. Dart claimed that injuries such as these were inflicted by rival apemen. Subsequent research has shown that they were made by predators, in this case a leopard that had dragged off the hominid in its mouth.

Born in 1903, the son of missionaries, Louis Leakey grew up among the East African Kikuyu people whom his parents were trying to convert, and at the age of 11 he was initiated as a member of the tribe. He studied anthropology at Cambridge and became convinced that early humans had indeed originated in Africa – much against the intellectual run of play of those days, as we have seen. And he became certain that the Olduvai Gorge – a steep-sided ravine about 30 miles (48 km) long and 300 ft (90 m) deep in the eastern Serengeti Plains of northern Tanzania – offered a unique opportunity for science to learn about our prehistory. "Olduvai is a fossil-hunter's dream," he once said. "It shears 300 feet through stratum after stratum of Earth's history as through a gigantic layer cake."[20]

However, he was so certain about man's African genesis that he made several early errors of judgment from which his reputation took a long time to recover. After one visit to the Olduvai Gorge, he announced that a skeleton found there by the German scientist Hans Reck was an ancient member of our lineage. In fact, it turned

THE BUSH OF LIFE

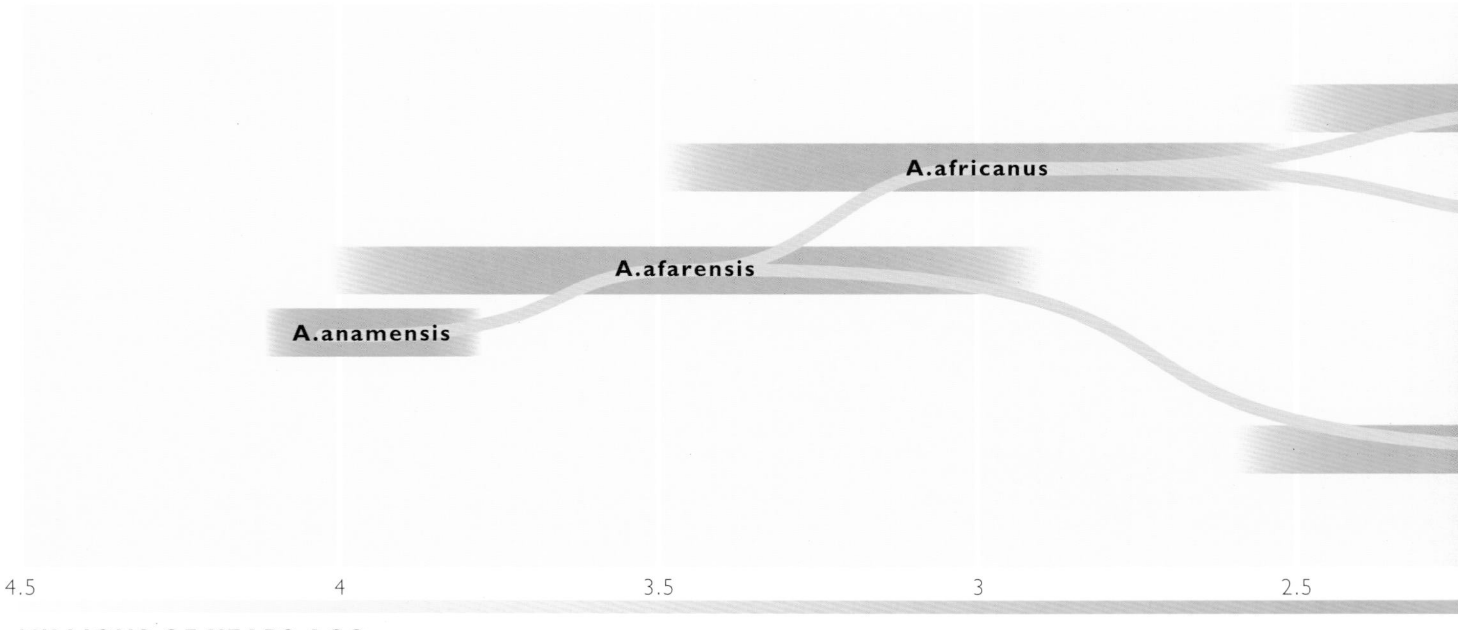

MILLIONS OF YEARS AGO

out to be a member of modern *Homo sapiens* that had been buried relatively recently in far older sediment beds – a common problem in paleontology, but not an insurmountable one. On another occasion, an eminent professor, P.G.H. Boswell, was due to visit one of Leakey's prime sites. Just before Boswell's arrival, Leakey discovered that the metal pins that he had used to mark the precise place where his most important fossils had been found had been stolen by a local fisherman to make hooks for his lines. Disgusted with Leakey for losing these sites, and with his general absent-mindedness, Boswell returned to Britain and began writing scathing papers about his work. This falling out with the establishment was only worsened when Leakey left his first wife, Frida, to live with Mary Nicol. This was simply not done in the 1930s, and the pair was ostracized.

Leakey and Nicol were a remarkable couple, however. She was a perfect match for him in terms of her unconventionality, her desire for adventure, and her ability. She smoked cigarettes, wore long pants, and could pilot a glider. She never completed

This chart depicts one of the most likely routes by which *Homo sapiens* evolved, although scientists disagree over the exact relationships between species and their common ancestors.

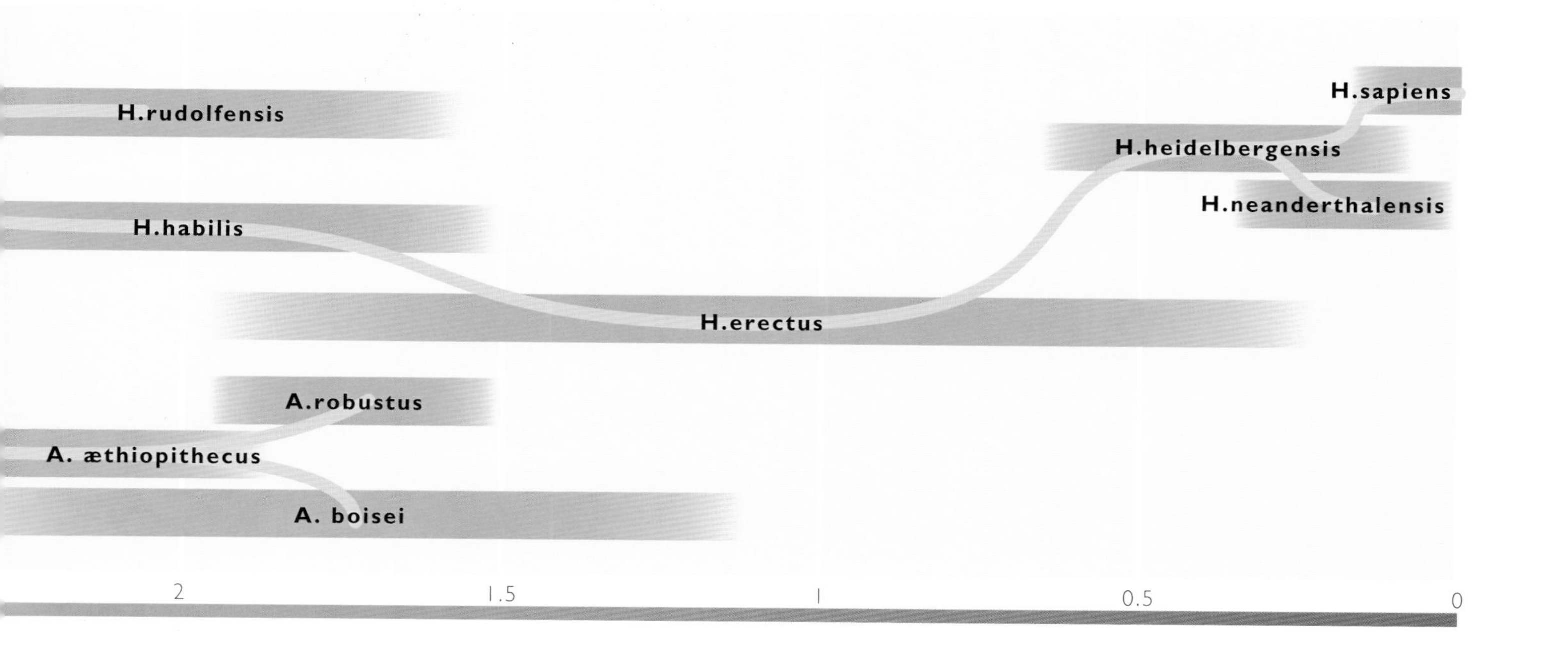

a school course, never sat for an examination, and never gained a single academic qualification, although she more than compensated for this lack later in life when she was given several honorary degrees. In spite of her nonacademic background, Mary had become obsessed with archaeology after a series of visits to Les Eyzies and several other prehistoric sites in France. (We shall encounter several of these later in this book.) Leakey and Nicol met at a dinner party, she did some drawings of stone tools for his book *Adam's Ancestors*, and they formed a clear, mutual attraction for each other. Her mother tried to block the romance (after all, Leakey was married and had two small children) by taking her wayward daughter off to South Africa. They got as far as Victoria Falls, where Mary left her mother and flew to Tanganyika (now Tanzania) to join Louis on his fourth East African expedition.[21] Here the pair began their joint, lifelong association with the Olduvai Gorge. Later they returned to London and got married.

However, Leakey found it extremely difficult to raise funds for his expeditions. He wrote an anthropological study of the Kikuyu, as well as other works, but could not find an academic post. At one point, after his return to Africa, he had to sell beads and beeswax to make money. Then, in 1939, he joined the Kenyan government as a civilian intelligence officer before being drafted into the African Intelligence

LEFT: Olduvai Gorge, where so many of the dramas of human evolution have been unearthed by scientists. BELOW: Zinjanthropus – now known as *Australopithecus boisei*. Louis Leakey initially claimed the species was a tool-making ancestor of modern humans, but later "downgraded" his discovery as an evolutionary offshoot. Louis Leakey discovered the first Zinj skull in 1959 at Olduvai Gorge.

THE PILTDOWN HOAX

In December 1912, British scientists revealed that they had made a truly sensational fossil discovery: fragments of skull and jaw of an ancient hominid. The find was made in a gravel pit near Piltdown Common in Sussex by amateur geologist Charles Dawson, and was analyzed and confirmed as belonging to a humanlike being of great antiquity by Arthur Smith Woodward, the leading British paleontologist of the time. According to his reconstruction, Piltdown Man had a large brain but possessed jaws and teeth very like an ape. The discovery was seized upon by the British scientific establishment. Many other nations – including France, Germany, and Indonesia – had already been revealed as possessing human fossil sites of great importance. Now Britain could also claim that key parts of the drama of human evolution had been played within its shores. In addition, Piltdown Man confirmed what many paleontologists then believed: that the human brain had evolved millions of years ago, before we became carnivores or upright apemen, and that intelligence had long ago differentiated mankind from the rest of

the animal kingdom. In fact, Piltdown Man turned out to be a hoax, as many European and American scientists had suggested not long after its unearthing. It had been put together from fragments of human skull and bits of orangutan jaw, all altered and stained to look like fossils. However, the fraud was not confirmed until 1953, when fluorine analysis carried out by dentist Alvan T. Marston showed that both the skull and the jaw were in fact of very recent origin. The perpetrator of the Piltdown Hoax was never uncovered, although Charles Dawson himself remains a favored candidate. However, its "discovery" was influential for several reasons, the most critical, and unfortunate, being the way it misdirected mainstream human paleontology, not just in Britain, but in many other countries. Finds – such as those of Raymond Dart – that indicated that our ancestors were actually small-brained even until relatively recently were discounted for decades because they did not fit into the fabric of scientific prejudice that had been endorsed by the finding of Piltdown Man.

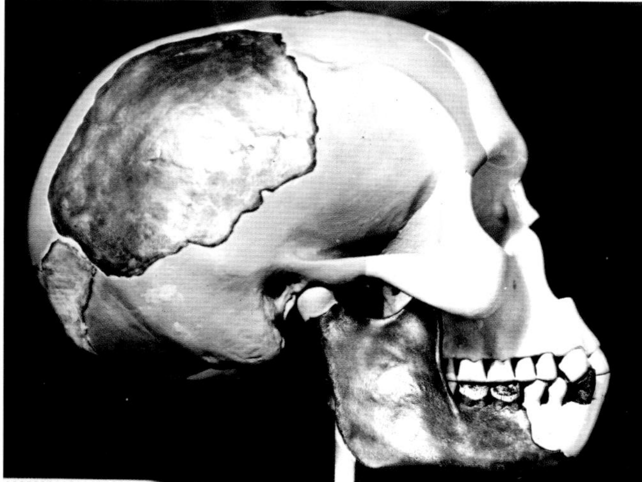

FAR LEFT: Excavations at Piltdown around 1913. Arthur Smith Woodward is shown second from left and Charles Dawson, Piltdown's discoverer, is on the right.
ABOVE: The Piltdown skull, constructed from a human skull and an orangutan jaw, fooled scientists for over 40 years.
LEFT: Alvan T. Marston, a British dentist who first suspected the hoax, at home with some of his anthropological specimens, 1953.

Department. During World War II, among many tasks, he helped supply arms to Ethiopian guerrillas fighting Italian troops who had occupied their country. Leakey was a gun-runner, in other words.[22]

After the war, the fortunes of Mary and Louis Leakey slowly improved and, by 1951, they were back at Olduvai excavating in often grueling conditions. At the end of each rainy season they would begin a hunt for new fossils uncovered by erosion. This task involved crawling up and down the sides of the gorge – eyes inches from the ground – stopping at the smallest fragment of promising bone or stone, which would be loosened and studied with a fine brush and a dental pick. All this in heat that sometimes reached 110°F (43°C). They continued to uncover hordes of stone tools of great antiquity, mixed in with the bones of a variety of animals, many of them now extinct, including giant pigs the size of rhinoceroses and large saber-toothed cats. Then, in 1959, came their breakthrough. Mary discovered a large piece of hominid skull and several teeth at a site surrounded by stone tools. She raced back to camp shouting, "I've got him! I've got him!" Louis, who was recovering from flu, returned with her. "We almost cried with sheer joy, each seized by that terrific emotion that comes rarely in life," he recalled.[23] Louis Leakey announced that they had found the earliest-known human ancestor of *Homo sapiens*, and he named it *Zinjanthropus boisei* – Boise's East African man. (Zinj is an ancient name for East Africa, while *boisei* honored Charles Boise, who had been funding much of the Leakeys' post-war work.)

Although australopithecines are not thought to have made tools, they may have used stones to smash open bones to extract marrow.

The skull – dated as being about 1.75 million years old – was subsequently revealed to have a distinctly odd anatomy. Running across its top was a massive bony ridge called a sagittal crest, down to which the powerful muscles of the lower jaw were attached. The face was curiously dished; the cheekbones flared like great flying buttresses; and the cheek teeth were enormous and the front teeth tiny. As Roger Lewin of the Peabody Museum at Harvard University puts it, "A most striking and bizarre physiognomy, to be sure."[24] Nevertheless, Leakey was convinced this was "the connecting link between the South African near-men (i.e. *africanus*) ... and true man as we know him." This was, he added, "the oldest well-established stone toolmaker ever found anywhere".[25] Just look at the stone tools they had made, he pointed out. And, certainly, the evidence of the creature's creative power lay scattered all around. Later named and categorized by Mary Leakey, these are the crudest hominid implements ever to be discovered and they represent the dawn of human culture. Most tools of the Oldowan industry (Oldowan is the adjective from Olduvai) consist of simple cobbles and cores of rock from which thin, sharp-edged flakes were struck. These flakes were probably used to slice open carcasses, a point graphically demonstrated by Louis Leakey, who, on several occasions, stalked gazelles, killed them with his hands, and then skinned them with one of the prehistoric tools. He also tried to skin animals, including rabbits and antelopes, with his teeth, but failed: hence the invention of stone tools, he surmised.

Pebble tools, about 1.7 million years old, found in the Olduvai Gorge. Such tools are the work of the earliest stone-tool makers, *Homo habilis*.

The question was: who had made these early implements? Leakey was certain it was *Zinjanthropus boisei*. After all, it was in the right place at the right time, he pointed out. Other paleontologists were not so sure. Zinj looked awfully like the robust australopithecines that had been dug up in South Africa – look at those great jaws for grinding vegetation. Surely this must have been a highly specialized creature, they said. In addition, no one had ever previously found a convincing implement associated with an australopithecine, for all Dart's grandiose claims about his "southern apes." Leakey remained adamant, until chance once more intervened. The Leakeys had always made a point of taking their sons – Jonathan, Richard, and Philip – on as many of their expeditions as possible, and tried to involve them in the work. This idea would pay rich dividends, both in the short term and in the long (as we shall see later in the book). In May 1960, the Leakeys' eldest son, Jonathan, then 19, was helping at Olduvai. He was wandering around the site when he spotted an odd fossil: the lower jaw of a saber-toothed cat. Such fossils were rare, and the find was deemed interesting enough to merit further investigation. Jonathan began his own excavation, and within days had struck fossil gold, unearthing several pieces of skull of a hominid child about 12 years old. Over the next three years, more and more hominid

A *Homo rudolfensis* skull (a close relative of *Homo habilis*), found in 1972 by Richard Leakey's team at East Turkana, Kenya. The skull was constructed from 150 pieces of bone and is surrounded by remaining fossil fragments. It is thought to be about 1.8 million years old.

fossils like this were found in Olduvai, mixed in with stone tools and animal bones. The partial remains of at least three individuals (including "Jonny's child", as Leakey Jnr's discovery was known) were eventually uncovered, along with ambiguous evidence of a crude campsite. Anthropologists speculate that the trio may have died and been left behind by the rest of their tribe, their corpses then being ripped and chewed up by hyenas.

But it was the nature of the fossils that caused the real surprise. These people had large brains (at least compared to any australopithecine), thin-boned, humanlike skulls, and teeth like a modern human's. They were far more likely candidates to be stone-tool makers than Zinj. In other words, the latter had probably been an intruder who had stumbled into foreign territory, where he met his fate. Leakey promptly abandoned his claims for *Zinjanthropus boisei*, and accepted what everyone else had been telling him: that Zinj was an australopithecine. And, indeed, that is its classification today: *Australopithecus boisei*, which we encountered earlier in this chapter. As for the new hominid, Leakey (in conjunction with his colleagues, paleontologist Philip Tobias and anatomist John Napier) assigned it the name *Homo habilis* – handyman – the name having been suggested by Raymond Dart. Leakey's critics complained this new classification was based on insufficient evidence – not more than a few pieces of fossil. More *habilis* remains have since been unearthed, however, and it is now generally accepted both as the creator of the Oldowan industry and as a direct ancestor of *Homo sapiens*.

Habilis is important to our story because its emergence marks the point at which our lineage turns away from the australopithecines. With the appearance of Leakey's handyman more than 2 million years ago, it seems that one branch of the hominid empire had managed – after 3 million years of evolution – to cross the cerebral Rubicon. It had started to make tools and its brain was beginning to expand – although not spectacularly. *Habilis*'s brain volume of around 700 cc is still only about half that of *Homo sapiens*, but it was about 50 percent more than *africanus*'s or any other australopithecine – which certainly made it the most likely maker of the stone tools that lay scattered around its remains.

So let us recap. *Afarensis* apemen evolve into *africanus* apemen, as well as into two species of large, vegetarian australopithecines: *robustus* in South Africa and *boisei* in East Africa. Then *africanus* disappears and *Homo habilis* appears, leaving us, about 2 million years ago, with two hominid experiments taking place side by side: vegetable-munching, big-jawed, relatively small-brained apemen (*boisei* and *robustus*); and larger-brained, tool-making, humanlike beings (*Homo habilis*), who must have been at least partially carnivorous (hence the tools). "These two creatures lived side by side [and] represented two different experiments toward what eventually became man," said Louis Leakey.[26]

It is a fairly attractive menu: *afarensis*, followed by *africanus* (with a couple of side orders – *robustus* and *boisei* – that you can leave on the table), finishing up with *Homo habilis*. Unfortunately, the study of human evolution is never that convenient. For example, some paleontologists think the line that led to *habilis* may have evolved directly out of *afarensis* stock, bypassing *africanus*. This means that *africanus* evolved only into *robustus* and *boisei* and was, therefore, an evolutionary dead end. Others point to the severe problems that have been encountered in investigating *Homo habilis*. One set of fossil remains (labeled OH 62) that are attributed to *habilis* were discovered by

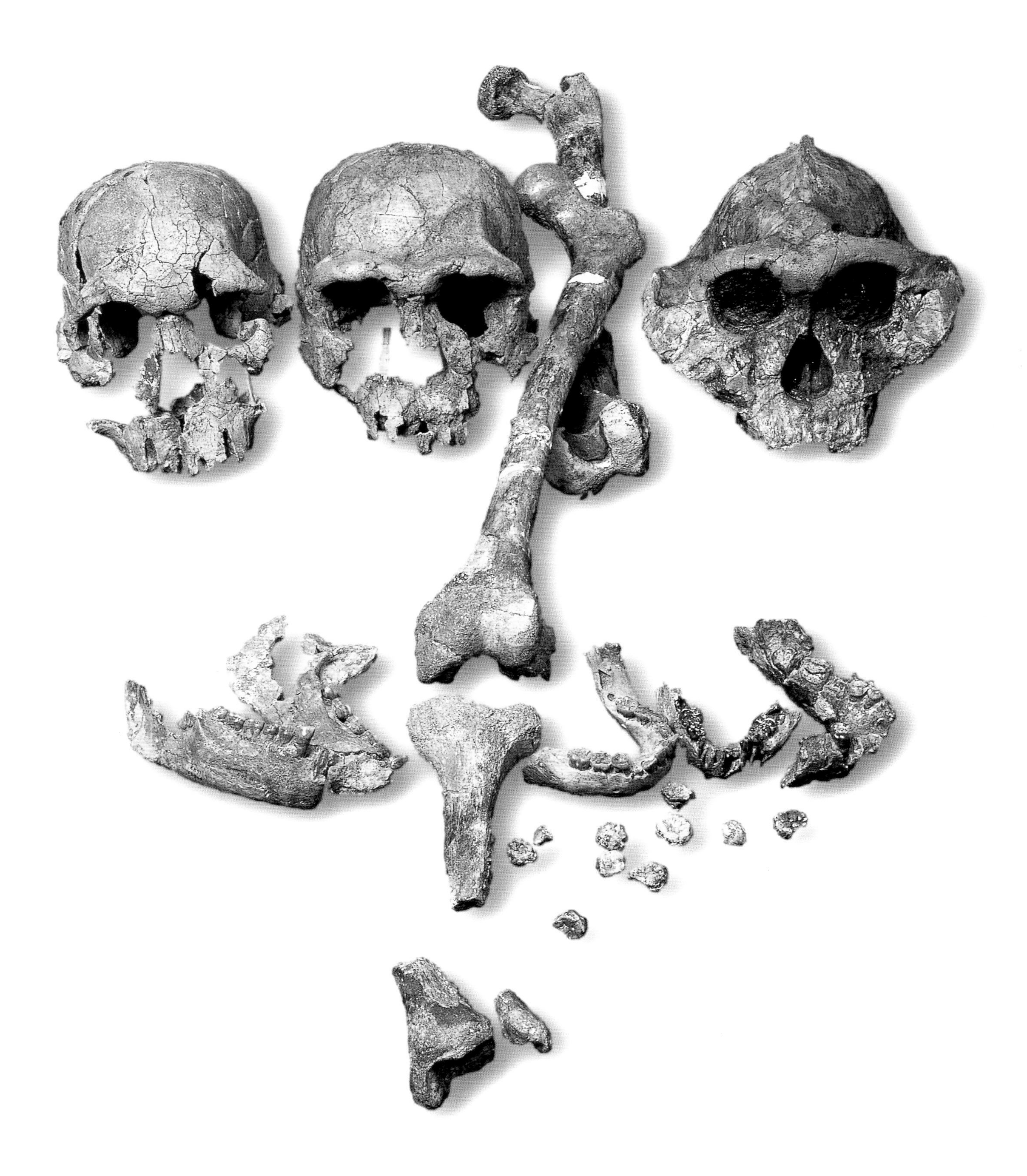

Don Johanson and Tim White in the Olduvai Gorge in 1986. The remains were important because they included upper and lower limbs – one of the first times that arms as well as legs had been obtained for a single *habilis* skeleton. However, the bones were hopelessly splintered, in Johanson's own words, and took a long time to analyze. When that was completed, the results caused surprise – to say the least. Far from being a larger, more *sapiens*-like hominid than its australopithecine predecessors, this *habilis* seemed to be more apelike than Lucy, with relatively long arms compared to its legs. Indeed, some scientists now refer to the species as *Australopithecus habilis*.[27]

Somehow, evolution seemed to be turning on its head. "I don't like *habilis* as a species; something is all wrong with it and always has been," says Alan Walker, a British paleontologist whose involvement in the unfolding story of human evolution will play a major part in the next part of our saga.[28] Meave Leakey is similarly unhappy about *habilis*'s role as a halfway house between australopithecines and our emerging lineage. "The trouble is that we lack good evidence. We desperately need a good skull and skeleton of a single member of the species to resolve this problem. Certainly this stage of human evolution is one of the most difficult to understand."[29] *Habilis*, in short, is a problem. It may turn out to be another evolutionary route that branched away and was destined to wither on our bush of life, while some other intermediary evolved from australopithecines and then formed the branch that led to the rest of the *Homo* pedigree. It is also possible that remains attributed to *habilis* may actually belong to more than one species. Several splitters – scientists such as Bernard Wood of George Washington University – believe that a second species of hominid, *Homo rudolfensis*, is responsible for some of the skulls and bones that have been attributed to *habilis*. This rather mysterious member of the human bush of life has a squarer jaw and a longer face than *habilis*.[30] Hominid evolution was a busy business, it would seem. The *Star Wars* bar was getting busier than a Glasgow pub on a Saturday night.

However, *habilis*'s shortcomings should not worry us unduly. Even though its status – as the first hominid species granted the honor of being a member of the *Homo* line – has been questioned, it is clear that an intellectual gulf had indeed been crossed by our ancestors more than 2 million years ago: witness those stone tools littered across the "living floors" of the Olduvai Gorge. Man the tool-maker had arrived. The exact identity of the species responsible may not yet be agreed, but its impact was considerable. Within a few hundred thousand years (by just under 2 million years before present) its successor, *Homo erectus*, manifested itself in profound and

OPPOSITE: A collection of hominid fossil skulls that clearly demonstrates the *Star Wars* bar theory. Discovered at East Turkana, Kenya, during expeditions led by Richard Leakey in the 1970s and 1980s, the species shown, from right, are: *Australopithecus boisei*, *Homo erectus*, and *Homo rudolfensis* (a close relative of *Homo habilis*), all of which coexisted in Africa between 2 and 1.5 million years ago.

dramatic ways, as we shall see in the next chapter. *Erectus* was definitely a tool-maker, a primitive mover and shaker. But before we move on, we have one last set of questions to consider. Why did one branch of the australopithecines, who had happily strode the woodland, forest, and savannas for so long, evolve into a new form of hominid? After 2 million years of quiet, upright success, one set of apemen finally developed into beings we now accept as humans. But why? After such an astonishingly long time of being upright apemen with hands free for creative work, why did our ancestors begin at last to make tools and start, very slowly, to evolve bigger brains?

The answer, perhaps not surprisingly, takes us back to the same forces that had shaped the australopithecines' path from the trees: climatic change. By about 2.5 million years ago, just as *Homo habilis* was beginning to take shape, the world's weather systems entered a period of intense fluctuation. Bouts of warming and cooling

Sparse forest clings to the edge of the Mara River in Kenya. This image reflects the effect that climatic change would have had on the evolution of our hominid ancestors. The open grassland would once have been entirely covered with forest. As increasingly hot conditions reduced forest areas, tree-living primates were forced to adapt to life on the open savanna.

AFRICA – MAJOR AUSTRALOPITHECINE FOSSIL FINDS

OLDUVAI GORGE: THE CRADLE OF HUMAN EVOLUTION

Olduvai Gorge is a 30-mile (50-km) long gash that runs through the Serengeti Plains of Tanzania, and it was here that Louis and Mary Leakey made virtually all their great discoveries of early human fossils. It is a parched, arid, uninviting place – although it was not always so inhospitable. Two million years ago, a lake occupied much of the region. Its water was highly alkaline, however – a result of ash that rained down from the nearby volcanoes of Kerimasi and Olmoti. This downpour had fortuitous consequences: shallow alkaline lakes can support more biomass than most other bodies of water. As a consequence, algae flourished, along with fish such as tilapia and birds such as flamingos – making Olduvai a magnet for other wildlife and predators, including humans.

However, the showers of ash, along with other geologic and climatic changes, eventually began to cause the lake to dry up, and by about 600,000 years ago it was reduced to a series of small ponds. Then, about 500,000 years ago, the main water course in Olduvai

changed direction and started to flow eastward, as it does today. The stream began to cut the gorge, exposing the layers of volcanic ash that had rained down there over the millennia, as well as the bones and tools that were embedded in them.

The Olduvai Gorge has been carefully surveyed by scientists, such as Richard Hay of the University of California at Berkeley, and the resulting geologic evidence has provided paleontologists with invaluable knowledge about the behavior of early men and women. For example, virtually all the rock sources from which stone tools were chipped have been identified, allowing researchers to trace humans' growing awareness of the suitability of different materials for making tools. Early sediment beds, those about 1.9 million years old, contain tools made of lava that came from only about a mile away. Those from more recent times reveal that a complex pattern of trade routes in many different types of rock had now been established by tool-makers.

LEFT: The Olduvai Gorge, a canyon cut into the south-eastern Serengeti Plains in Tanzania. The renowned paleontologists Louis and Mary Leakey spent many years working in the Olduvai Gorge, and many of their great hominid fossil discoveries were made here.

ABOVE: Examples of stone tools discovered at Olduvai Gorge. These implements are typical of the Oldowan industry – simple pieces of rock from which sharp-edged flakes were struck. They are thought to have been made by *Homo habilis* about 1.7 million years ago.

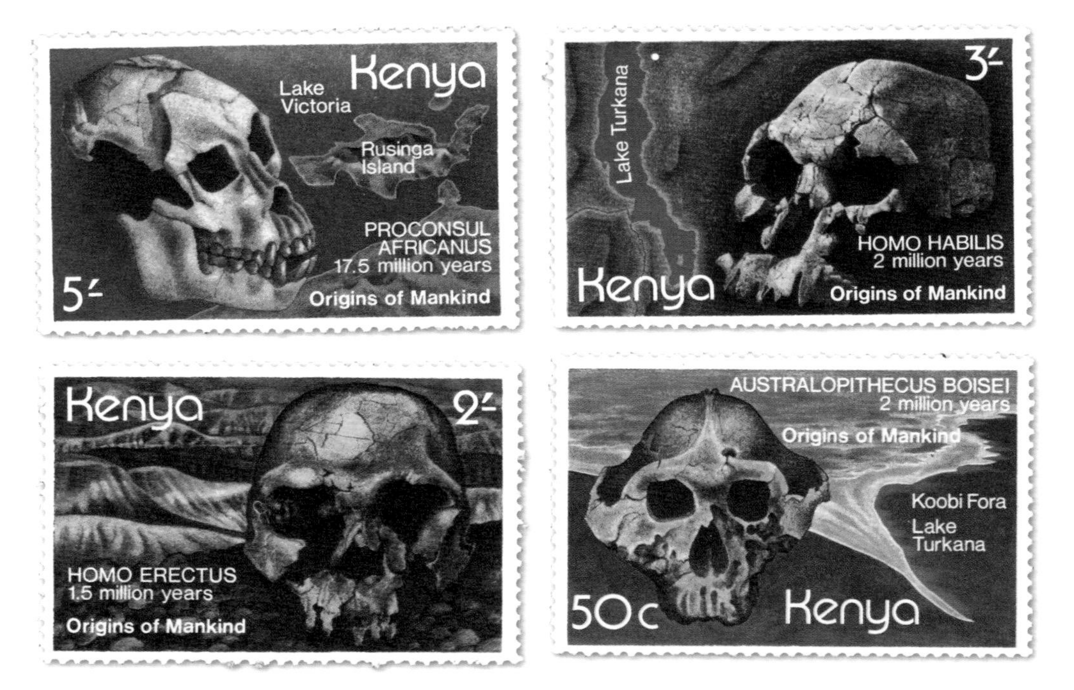

The "Origins of Mankind" series of stamps produced in Kenya to commemorate the magnificent fossil discoveries made within the country's borders.

occurred every few thousand years, while ice sheets began a stately pavan of advance and retreat from the planet's poles. Ocean currents shifted and realigned, sometimes in less than a decade. "The abrupt coolings most likely devastated the ecosystems on which our ancestors depended," says neurophysiologist William Calvin of the University of Washington School of Medicine. "Because of lower temperatures and less rainfall, the forests in Africa dried up and animal populations began to crash. The progenitors of modern humans lived through hundreds of such episodes, but each was a population bottleneck that eliminated most of their relatives. We are the improbable descendants who survived."[31]

But how? What lucky trick did our ancestors possess, what knack for survival did they have? The answer was increased improvisation and improved versatility in behavior. Our ancestors were the ones who chose to eat meat as well as vegetables. We became more and more omnivorous. The herbivore aspect was easy – that was part of our primate ancestry – but the carnivore portion was trickier. That required a new gambit, as Richard Potts of the Human Origins Program at the Smithsonian Institution in Washington, makes clear. "About 2.5 million years ago, hominids encountered even greater fluctuations in the climate," he says. "At the same time we

Richard Leakey in the vault of the National Museum of Kenya, with his *Star Wars* bar of hominid fossil finds (see page 60).

see the appearance of stone tools. That is no coincidence. They indicate that at least one species of hominid was responding to these changes by becoming even more adaptable, rather than becoming specialized in the way that *robustus* and *boisei* did. By making tools, dietary choices became even greater. Not only could people skin the large, dead, and undoubtedly smelly carcasses they occasionally found, they could crack open their bones for marrow. In addition, tools would also have helped pound and break down vegetables and nuts that could otherwise only have been eaten by animals with specialized dentures, and also help dig up tubers which are rich in protein and calories. Just as australopithecines responded to oscillating climates by walking, by becoming more versatile movers, so did the first members of the *Homo* line 2 million years later. They made tools and became more versatile eaters."[32]

The first stone tools appeared about 2.5 million years ago and were, as we have seen, fairly crude. Nevertheless, it is clear that early humans came to rely on them more and more, for these implements are found farther and farther from the sites at which they were manufactured. "We became less tethered to our habitats," as Potts puts it. Our ancestors' behavior was becoming increasingly diverse, our menus more adventurous. Mankind was on the move.

CHAPTER THREE

MASTER OF THE SAVANNA

In a dark, cool room in a white, concrete, 1970s-style building in downtown Nairobi, a series of foam-lined wooden boxes are stored inside metal cabinets that line the walls of this mysterious sanctum. Neatly typed on one side of each box is a group of letters followed by a number. The air outside is noisy, hot, and thick with the fumes of city traffic. Within the room, however, there is an aura of reverential silence and air-conditioned calm. This is the vault, and it is the holiest place in the National Museum of Kenya, a bomb-proof chamber where the bones of mankind's ancestors have been assembled and carefully stored for the edification of future generations. There is only one door, made of 9-inch-thick metal, and it opens into a room that is – effectively – a bank of mankind. Bits of australopithecines, fragments of *Homo habilis*, and portions of their successor, *Homo erectus*, are all preserved with care and precision. Many of the vault's deposits contain only one or two fossil scraps, representing the pitifully few pieces of a long-extinct species that scientists had been able to gather from the arid soil of Kenya's desert heartlands. A couple of boxes contain up to a dozen morsels of bone, enough for scientists to glean a great deal of important data about a species' brain size or its ability to walk upright. However, there is one fossil array that quite dwarfs the rest. Its components are so numerous – a total of 106 bones – that they fill an entire cabinet. It is a staggering assemblage, given the fossils' antiquity. These are the remains of the Nariokotome boy.

Richard Leakey with a 1.6-million-year-old *Homo erectus* skull unearthed from Lake Turkana in 1984.

The boy, we now know, was a member of *Homo erectus*, a species that was first discovered in 1887 but remained a relatively enigmatic member of our lineage, our knowledge of its anatomy restricted to only a handful of skulls and a few splinters of bone that had been found by paleontologists. In each case, the skull had thick walls and the braincase was reinforced by bony ridges across the back, top, and sides; the eye sockets were overshadowed by a "brow ridge" – a prominent ridge of bone that gave the skull a glowering expression; and there was a low, receding forehead leading to a long and flat crown. The teeth were distinctly smaller than those of australopithecines and *habilis*, but the lower jaw was still thick-boned and chinless. As for the species' demeanor, deportment, or size, scientists had little clue. *Home erectus* was a mystery – until August 1984, when a scientific team led by Richard Leakey, and which included the noted English paleontologist Alan Walker, stumbled on the Nariokotome boy, a discovery now rated as one of the most outstanding in human evolutionary studies.

Leakey – who had by now taken on the role of fossil-hunter from his father Louis, who died of a heart attack on October 1, 1972 – was working with Walker in the Nariokotome region on the west bank of Lake Turkana, Kenya. The expedition included half a dozen gifted Kenyans from the Wakamba people. They had been trained by Leakey to find fossils and had gained such a reputation for their prowess that they became known as the Hominid Gang. Strolling in the searing heat, they could spot a tiny speck of fossil against the gray desert background of Kenya's parched soil with unerring acuity. And the most proficient of all these bone-hunters was Kamoya Kimeu, the gang's foreman. As Meave Leakey – head of paleontology at the National Museum of Kenya and Richard's wife – puts it, "No one can find them like Kamoya."[1]

Kamoya Kimeu, the eagle-eyed member of the Hominid Gang renowned for his ability to spot even the tiniest fossil fragments.

As he wandered along the dry bed of the Nariokotome River one day, Kamoya found a sliver of human skull. "It was the size of a matchbook and the color of pebbles," recalls Walker. "Lord knows how he saw it."[2] Leakey recognized the fragment as belonging to *Homo erectus* but was unimpressed. "Seldom have I seen anything less hopeful," he recorded in his field diary.[3] Walker was equally pessimistic. "Our hearts sank when we saw the small fossil, a rectangular piece about one inch by two. We had followed up a hundred scrappy finds like this and found nothing more."[4] So, the next day, he and Leakey went off to enjoy other paleontological pleasures while Kimeu and the rest of the Hominid Gang began their search in earnest. First they used wooden staves with 6-inch nails hammered into them to disturb the site's top layers of soil. Then they scooped up the dirt and shook it through wire-mesh sieves to separate the tiny grains of soil and stone in the hope that these would leave behind any larger bones. It is one of the most tedious jobs in paleontology, as Leakey acknowledges. "We all try to find urgent business elsewhere when sieving time comes," he once admitted.[5] On this occasion, however, Kimeu and the rest of the gang hit pay dirt – for when Leakey and Walker returned to the excavation in the evening they found that the Hominid Gang had unearthed many more segments of *Homo erectus*: a forehead, part of the skull that surrounds the right ear, and the right and left parietals, which form part of the vaulting of the braincase. Scepticism was transformed to delight, and a special camp dinner was served that night: smoked fish, tomato soup, steak, cheese, Beaujolais, coffee, and port.[6]

The team continued to excavate and was quickly rewarded with more bones, and very special ones at that: hominid ribs. Walker was incredulous. "We never find hominid ribs," he recalls in *Wisdom of the Bones*, which he coauthored with Pat Shipman. "Up until the time people started burying other people, a mere 100,000 or so years ago, ribs were among the first body parts to be crunched and munched into splinters by carnivores."[7] Nevertheless, the expedition had found ribs. Nor did their good fortune peter out: an entire skull, pelvis, legs, and an arm were uncovered over the next four weeks. Only the hands and feet of this long-dead ancestor eluded the team. "Taking shape before our eyes was an essentially complete *Homo erectus* individual, the first to be unearthed since this species was discovered almost a century ago," recalls Leakey. "It was an extraordinary experience for me, for all of us. Three

A recreation of the last moments of the Nariokotome boy, who died 1.5 million years ago – a victim, most probably, of blood poisoning, scientists believe.

weeks of paleontological bliss."[8] Walker was equally ecstatic – and awed. "Our scientific ancestors had spent their lives, expended their funds, risked their health, built and sometimes derailed their careers, all in the frustrating search for the missing link – and we had found it."[9]

The Nariokotome boy was a member of the *Homo erectus* species, as we have said, although some scientists now assign it to a post-*habilis* ancestral species of *erectus* that they call *Homo ergaster*. (Most still call it *Homo erectus*, however.) Studies of the sediments around where its skeleton was found suggested the find was 1.5 million years old; investigation of the pelvis showed it to be that of a male; and analysis of his teeth – by Holly Smith, an anthropologist at the University of Michigan – revealed these were the remains of an adolescent, probably 9–11 years old. His second molars, for example, had begun to wear down, but his wisdom teeth had only just started to form (although the rest of his skeleton was more like that of a boy nearer 14 or 15).

Indeed, it was from studying the Nariokotome boy's teeth that scientists gleaned clues to the most poignant aspect of the young lad's story: his fate. The shedding of one of his milk teeth seems to have triggered septicemia (blood poisoning). Without antibiotics, which protect us today from such infections, he would have been highly vulnerable to this condition. Thus he perished, falling face down in a marsh, where his body gently spread out as it rotted and was then buried under layers of swamp mud. He remained buried for 1.5 million years, until the covering sediments were eroded and some of the fragments were revealed to the ever-vigilant eye of Kamoya Kimeu.[10]

Members of the Hominid Gang sieving sediments – "the most tedious job in paleontology," according to Richard Leakey, although an essential part of the fossil-hunting process.

Although all the key parts of the Nariokotome boy were found within a month of that encounter, it took a further four seasons' excavating before Leakey and Walker could be sure they had siphoned up every shard of their precious acquisition. By the end of 1988, it was time to start thinking seriously about analyzing the bones in detail. The boy's remains were dispatched for storage at the vault in Nairobi, under the custody of Meave Leakey. Given that they were dealing with "a hauntingly complete" skeleton, as Walker describes it, paleontologists were confident it would provide exciting new information.[11] As we have seen, *Homo erectus* was most probably a

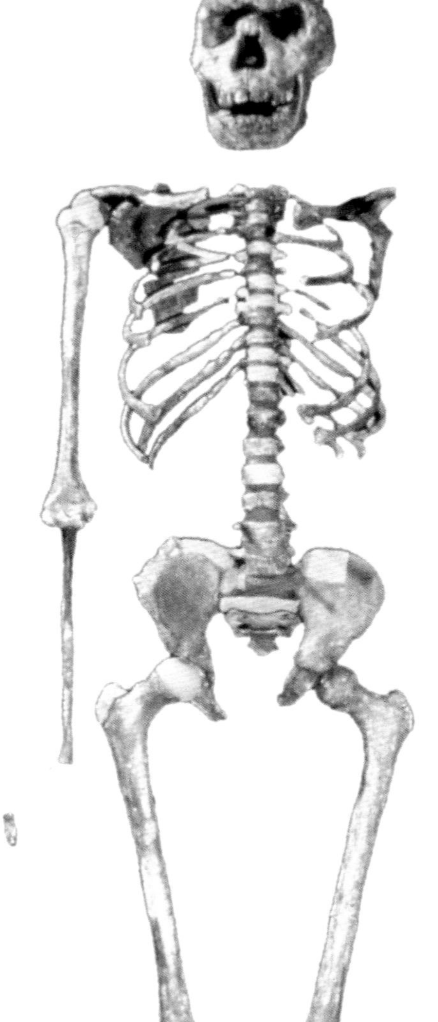

The skeleton of the Nariokotome boy: the first, and to date only, full skeleton of a *Homo erectus* unearthed by scientists. Discovered in 1984 by Kamoya Kimeu, Richard Leakey, and Alan Walker, the remains lack only hands and feet.

descendant of *Homo habilis,* which, in turn, evolved from australopithecine *africanus* stock. It was from this branch – *africanus* to *habilis* to *erectus* – that the bush of humanity's diverse forms was to grow, one of the twigs eventually flowering to produce *Homo sapiens*. Understanding the anatomy of *Homo erectus* was going to be critical to understanding that part of our evolution, scientists thought. And now they had an entire individual to study.

One of the first surprises for the team was the realization that the Nariokotome boy's body shape and size were very similar to those of modern people living in hot, arid places – tall, long-legged, and narrow-hipped. The discovery dramatically confirmed the antiquity of the human form, Leakey and Walker observed in *National Geographic*. "Suitably clothed and with a cap to obscure his low forehead and beetle brow, he would probably go unnoticed in a crowd today."[12] Estimates suggest the Nariokotome boy was about 5 ft 3 in (1.6 m) tall at death, quite impressive for an 11-year-old. If he had reached adulthood, he would have been an imposing 6 ft 1 in (1.9 m) or more. Far from being brutish and short, our predecessor was tall, leggy, and muscular. He also appears to have been well fed, as far as we can determine from his sturdy skeleton. The boy was probably about 78 lb (35 kg) at death and would have weighed nearly 154 lb (69 kg) if he had made it to adulthood. "What sort of prehistoric basketball player had we excavated at Nariokotome?" wondered Alan Walker.[13]

A tall, robust human was certainly not what had been anticipated. From the few meager leg bones of *Homo erectus* that had previously been unearthed, and from the fact that *erectus*'s australopithecine predecessors were small, perhaps no more than 4 ft 6 in (1.4 m) high (remember those bantam-proportioned hominids who left their trail over the Laetoli sediment beds), paleontologists had expected to confirm the fact that the species was modestly sized. Instead, they had unearthed a human being of astounding stature, contradicting "a long-held idea that humans have grown larger over the millennia," as Leakey and Walker put it.[14] Indeed, if ever a discovery demolished the canonical representation of human evolution as a stately linear progress to *Homo sapiens*, this was it. Remember those images of shuffling hominids, growing in stature as they advance irrevocably toward that pinnacle of biological development, modern humans? Well, compared to members of *Homo erectus*, we are the scuffling wimps, while they look like the towering masters of the savanna. It certainly makes a mockery of the notion that human evolution was a business of unswerving ascent to modern mankind.

Unlike his australopithecine ancestors, the Nariokotome boy and his kin were smooth-skinned. This adaptation, combined with his great height, meant that he had a large skin surface from which to perspire and cool down – essential for an active lifestyle in the arid conditions of the African savanna.

But what had been going on in the million years that led from *afarensis* to *erectus*? What had shaped the latter so that it now had such a startling aspect? One clue lay in the geometry of the boy's bones: perhaps his tall frame had evolved to give a large skin surface area that would assist heat loss in a hot, dry climate by sweating. We can see this effect today: populations have bodies suited to their places of origin. The Masai of Kenya tend to have tall, slender bodies, giving them a large surface area of skin from which to lose heat through perspiration. The adaptation is evident in the greater relative length of body extremities, especially the lower leg compared with the upper leg, and the forearm compared with the upper arm. The opposite proportions are found in cold-adapted peoples, such as the Lapps and Eskimos. In other words, the proportions of the leg and arm bones can act as a kind of thermometer, roughly indicating the average temperature of the land in which a population originated.

When measurements of the Nariokotome boy's arms and legs were fed into this "limb thermometer," the results suggested he was not simply of tropical origin – he

The shores of Lake Turkana in Kenya, where the bones of the Nariokotome boy were found.

was, as Walker puts it, "hypertropical."[15] In other words, he had evolved to lose heat with maximum efficiency, and that, in turn, implied that he must have been an extremely active individual who was capable of running around in the open, hot savanna. "He was a member of a species that – like later Englishmen – went out in the midday sun," says Walker.[16] And that certainly means he sweated. If he had not perspired, he could not have performed athletically. Perspiring is a key means for expelling heat in humans; without it, we would run the risk of suffering fatal heatstroke. We excrete warm water through our skin and it evaporates, thereby keeping our bodies cool – a process that is responsible for 15 percent of our daily heat loss.

And if the Nariokotome boy sweated, which is almost certainly the case, he had probably lost the body fur or hair of his more primitive ancestors, a covering that would have restricted his ability to expel heat. But without fur to shield him

LAKE TURKANA - MAJOR FOSSIL FINDS

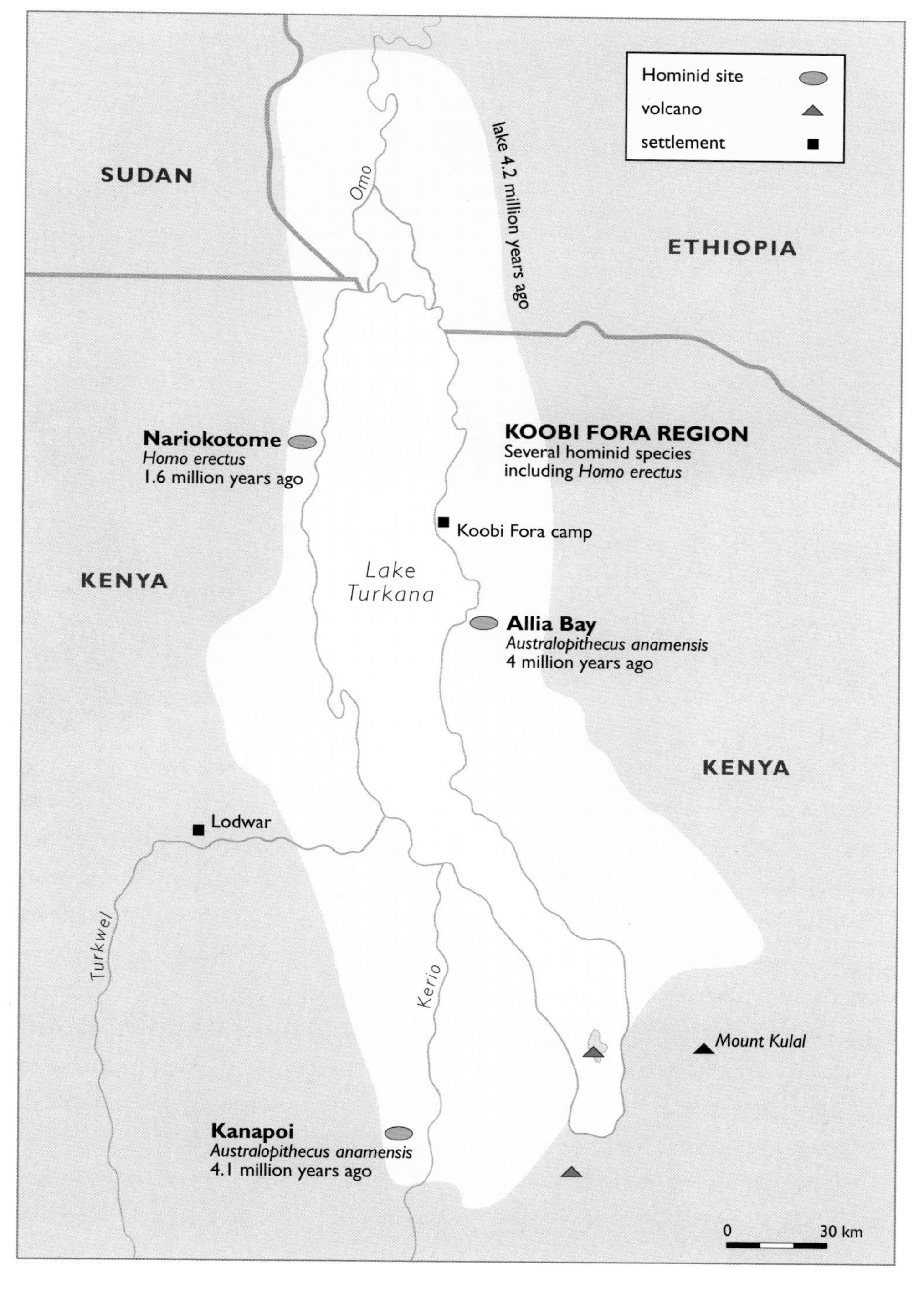

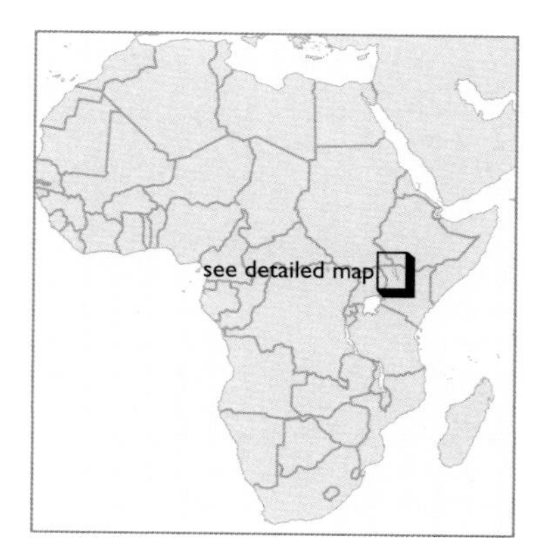

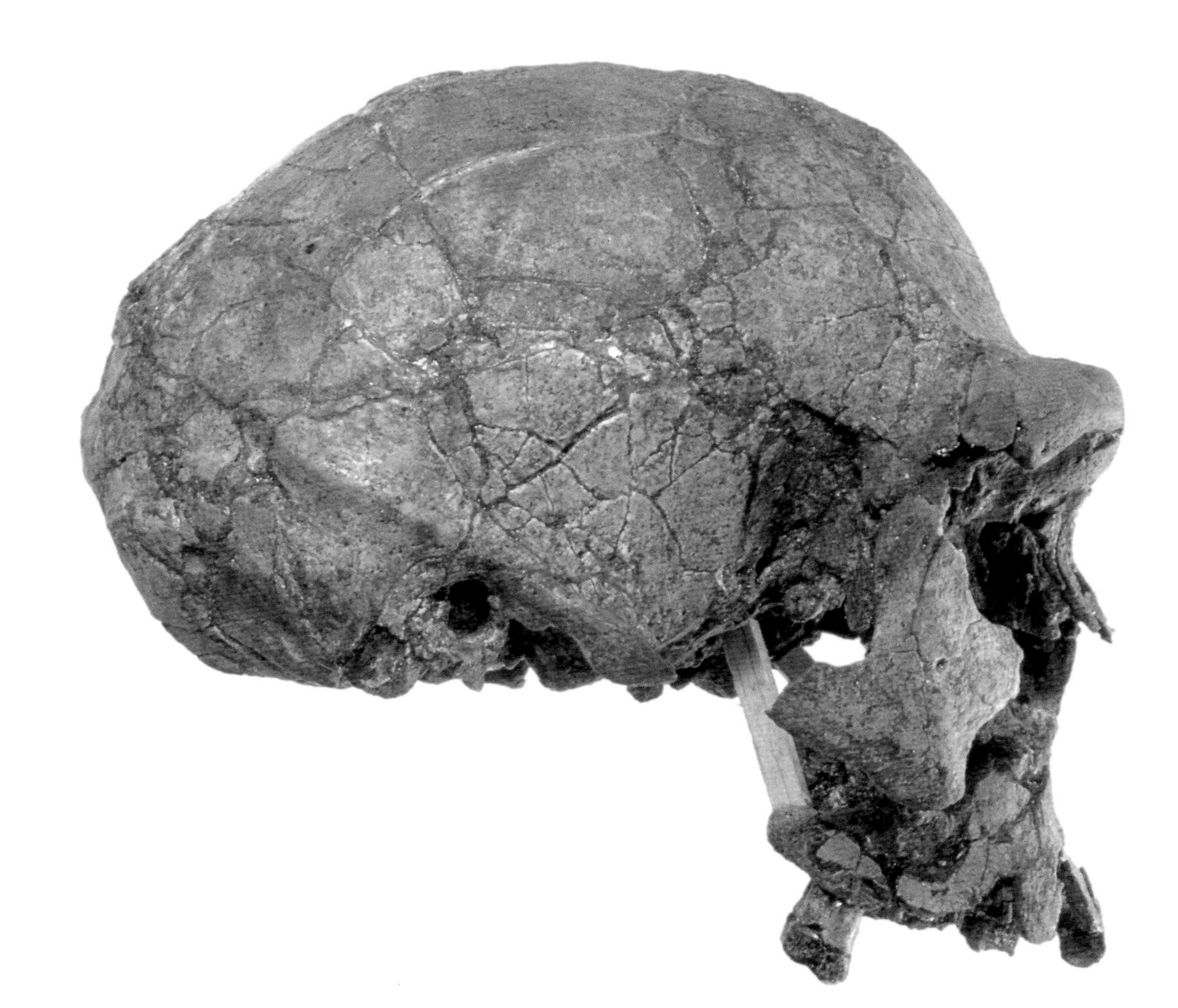

A *Homo erectus* skull known as specimen KNM-ER 3733, which was found at Koobi Fora in northern Kenya. Note the long, low skull and prominent browridge over the eyes, which is typical of the species.

from the sun's harmful ultraviolet radiation, he would have been dangerously unprotected – unless he had evolved a shielding of dark skin pigmentation. This point is stressed by Leakey. In *Homo erectus*, the thick body hair that had covered our more distant ancestors was reduced to fine, short body hair, he says. "At the same time ... our sweat glands developed ... a response to hunting on the open plains, where keeping cool became more of a problem than it had been for our ancestors. Once this covering of hair had disappeared, a dark-colored skin was a biological necessity."[17]

A picture was slowly emerging of this long-dead individual: tall, thin, smooth-skinned, and dark-colored. He was also exceptionally sturdy. Studies of his leg bones indicate that *erectus* must have been singularly robust; the species was built to withstand the stresses of a physically active lifestyle much more grueling than that of any hunter-gatherer tribes we know of today. This was a person designed for unusually intense exercise, for chasing and pursuing animals, or perhaps for avoiding enraged carnivores, in the midst of the hot African day.

Studies of the Nariokotome boy also showed that he had evolved an attribute that we seldom think of as defining our species. He had a nose. As anthropologists Bob Franciscus and Erik Trinkaus have pointed out, australopithecines had flat nasal openings, just like those of other primates. A protruding nose meant that humans could retain more moisture in their breath as they exhaled it, helping maintain an internal hydrothermal balance. It would have improved the boy's ability to conserve respiratory humidity and enhance his power to be active, in other words. As Trinkaus – who is Professor of Anthropology at the Washington University in St Louis – says: "There was a major change with the emergence of *Homo erectus* with respect to thermal and body fluid physiology, of which the shift in nasal morphology is one part and the body proportions are another. Our relative hairlessness – to be precise our short body hairs – and sweatiness probably emerged at the same time."[18]

However, it was the study of the Nariokotome boy's head and upper parts that were to provide the most notable and startling results. Initial measurements showed that his brain must have had a volume of only 880 cc – about double the volume of a typical ape and about two-thirds of the modern human average of 1350 cc. It may sound like a fairly respectable cranial capacity, but Walker disagrees. Compared with modern humans, the Nariokotome boy "had the height of a 15-year-old with the brain of a one-year-old." Not quite so impressive, in other words.[19] Other aspects of his skull's anatomy, on the other hand, did suggest a closer affinity with ourselves. In particular, his braincase had different-shaped right and left sides, which in people today is correlated with right-handedness.

But the real shock came when scientists studied the Nariokotome boy's spine. Ann MacLarnon, an anthropologist at the Roehampton Institute in London, found that the vertebral canal – a hollow passage in the spine that carries nerves from the brain to the rest of the body – was narrower near the chest than it is in modern humans. Together with other evidence, this suggested that *Homo erectus* did not have fine control over the muscles involved in breathing – muscles that we employ, quite unconsciously, in speech. Language, as we understand it, had probably not yet evolved.

Walker makes much of this feature. The Nariokotome boy, and all of his *Homo erectus* kith and kin, may have been the cleverest yet to walk on Earth – powerful stalkers who walked on two legs and who cared for each other – but there was

EARTH'S RESTLESS CLIMATE

Our planet's weather systems have never been stable. Myriad factors – tiny variations in the Earth's orbit, the inexorable movements of continents, and many other changes – have produced profound climatic oscillations, some of the most violent having occurred in the past 2 or 3 million years. Consequently, grasslands and forests have expanded and contracted, sea levels have risen and fallen, and land bridges between continents have emerged and sunk. Most scientists now accept that such fluctuations must have played a key role in human evolution, and understanding past patterns of climatic behavior is now shedding important light on the emergence of *Homo sapiens*.

One invaluable technique involves the study of atmospheric gases, especially oxygen, which comes in two principal varieties called isotopes. Both behave the same way chemically, but one isotope, known as oxygen-18, is slightly heavier than the other, which is called oxygen-16. Crucially, during warm periods, the latter evaporates from the sea slightly more easily than the former, while during cold periods more of the lighter isotope is stored in glaciers compared to the heavier. By analyzing the isotopic composition of marine fossils (which absorb oxygen to form their shells of calcium carbonate), or by probing the composition of ancient ice layers deep below the Antarctic surface, researchers can discover if particular sediments are rich or poor in oxygen-16. This reveals whether they were put down during warm or cold periods, thus helping build up a detailed picture of Earth's turbulent climatic past.

A scientist carves a core of ice that has been drilled from the Antarctic depths. Its oxygen content will provide vital clues to the planet's temperature fluctuations over the eons.

MAJOR EVENTS

- *Homo sapiens* first appears outside Africa.
- Emergence of *Homo neanderthalensis* in Europe (see chapter 6).
- Emergence of *Homo heidelbergensis* (see chapter 5).
- First well-dated occupation of western Europe by human-like settlers.
- Extinction of robust australopithecines in Africa.
- Earliest evidence of *Homo erectus* in Java.
- Emergence of *Homo erectus* in Africa.

CLIMATE VARIATION

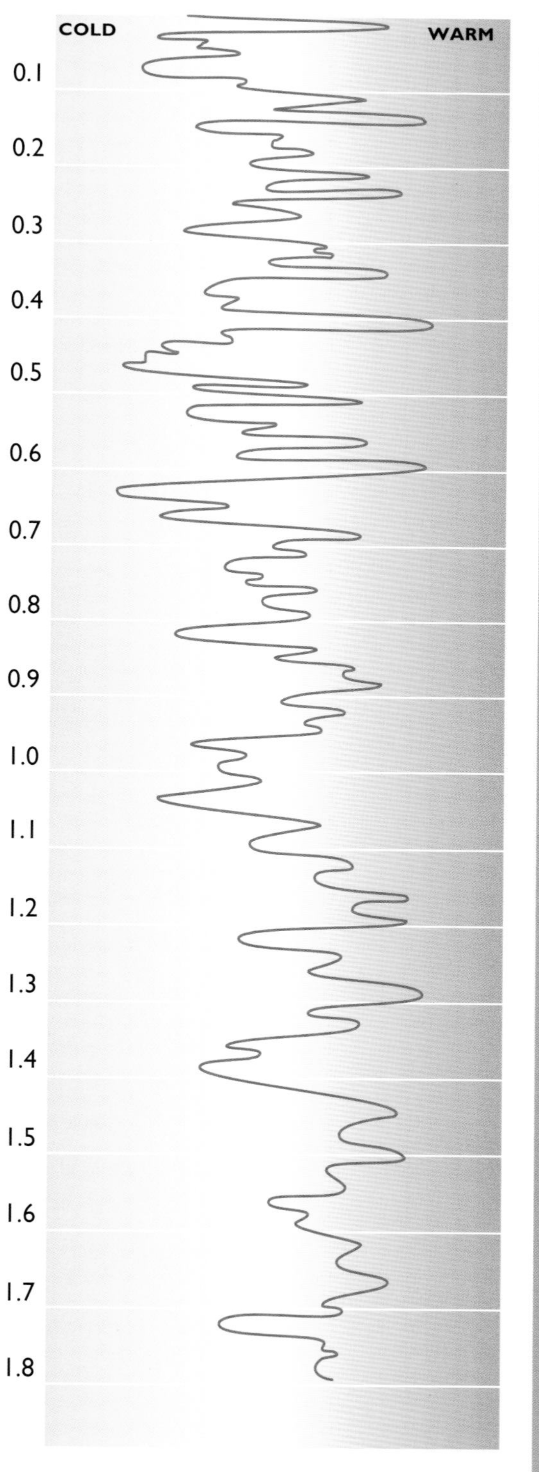

something missing. He had not mastered language, and that meant he was far from possessing what we understand as "human spirit". As Walker puts it, "In his eyes was not the expectant reserve of a stranger but that deadly unknowing I have seen in a lion's blank yellow eyes. He may have been our ancestor, but there was no human consciousness within the human body. He was not one of us." Starved of intellect, the Nariokotome boy could not talk, and he could only have been able to make crude mental maps of his world. "The boy my colleagues and I spent so many years discovering and analyzing was profoundly *in*-human," adds Walker. "He was large, he was strong, he was a tool-maker, a hunter, and an intensely social animal adapted to a rigorous, active life in the tropics. But he was not human, did not think like a human, and could not speak."[20]

A thoracic vertebrae of the Nariokotome boy (left, brown) and that of a modern human (right, white), showing the spinal canal. The space for the spinal cord in the former is half that of the latter, which suggests that the boy's power to control his breathing and, therefore, his speech, may have been limited.

This denial of *Homo erectus*'s human status is startling. The very fact that the species had been designated a member of the *Homo* lineage indicates that scientists had always viewed its members as being the first true men and women. Walker's analysis of the Nariokotome boy rejects this ranking. After waiting so long to find this missing link, one of its discoverers and chief investigators was now downgrading its importance, at least in terms of its intelligence and linguistic abilities. And don't forget the implements that *erectus* made. For more than a million years after the Nariokotome boy died, his species continued to make the same equipment – the Acheulian stone tool kit, as it is known, with its scrapers, axes, and cleavers. It was extraordinarily repetitive by any standards, and certainly does not speak of a species in the midst of an intellectual expansion. As Desmond Clark of the University of

California at Berkeley put it, "If these ancient people were talking to each other, they were saying the same thing over and over again."[21]

For her part, MacLarnon agrees that the Nariokotome boy's anatomy would have greatly restricted the control of his breathing, making long sentences an impossibility for him. It was MacLarnon's analysis of the spine that led Walker to deduce that the boy lacked speech. She compared the width of the spinal column and vertebral canals of a range of primates, including chimps and modern humans, and concluded that narrowness of the vertebral canal behind the boy's ribcage would have prevented him from controlling his breathing properly. "Lungs act like bellows, and spines carry the nerves that control the muscles which work those bellows," she says. "In humans those controls are very sophisticated – which allows us to enunciate without a break for long periods. We can utter lengthy, complex sentences and still have enough breath to emphasize key words, even though we have been speaking for a considerable length of time. We can give different intonations to phrases and get over extremely complex ideas when we talk. To judge from his spine, the Nariokotome boy just did not have the power to do that. However, that doesn't mean that he was incapable of making meaningful utterances. You can communicate a lot just by making simple grunts and sounds. But he couldn't have produced anything like modern speech."[22]

Such an analysis rests on two basic assumptions, however: first, that the boy is typical of the species, and second, that his bones were not damaged by disease or injury. In the first case, scientists can only assume his skeleton is characteristic of *Homo erectus* because it is the only complete one we have. Other *erectus* fossils – mainly made up of bits of skull – do seem to be very like those of the Nariokotome boy, supporting the idea that he was a fairly ordinary member of his kind. The second assumption is very different, however. His remains, including the spine, could indeed have been damaged by some kind of illness, and the effects could be skewing MacLarnon's measurements. This point seriously worries some academics, as highlighted in one recent study that indicates that the Nariokotome boy may have suffered from scoliosis, a curvature of the spine. "If that is the case, then obviously it would affect my data, but exactly how we cannot yet be sure," says MacLarnon. "In any case, since carrying out my studies of the Nariokotome boy, I have measured three other ancient hominid spines – all australopithecines – and have found precisely the same narrowness of the upper spine canal in the ribcage region. They must have been

poor speakers as well, and that provides support for the idea that their immediate descendant, *Homo erectus*, was similarly deficient. However, I admit that the jury is still out over the issue."[23] Walker, for his part, remains confident about the interpretation. "We can't assess easily whether or not scoliosis would have affected the boy's spinal canal, but the experts that I have spoken to say it would not have. All this does is make it easier for sceptics to dismiss our evidence."[24]

However, it is a considerable jump in logic to say that because *Homo erectus* could not articulate with the precision and clarity of an early-day John Gielgud, he must have been an "awful klutz," as Walker puts it.[25] He may have been a poor speaker, but that does not necessarily reduce him to a dumb animal capable only of mute, blank stares. This point is made by anthropologist Leslie Aiello of University College, London. "Walker uses the analogy that the Nariokotome boy has a brain the size of a one-year-old modern human, but all you need to do is turn this issue on its head and point out that the boy had a brain that was twice the size of a chimp's. And you don't see a blank yellow stare when you look into a chimp's eyes. You see an animal that has an unsettling affinity with ourselves. And this lad was twice as brainy as that, and much closer to us in evolutionary terms. It is true he probably did not have spoken language as we know it – the development of syntax is very important and probably did not come until later – but I would put money on the fact that he didn't have a blank animal stare. We would have recognized him as a fellow human being."[26] Her argument is backed by Meave Leakey. "*Homo erectus* was well on the way to being 'human.' Self-consciousness, compassion, and the emotions we commonly feel – I think these were all part of the *erectus* makeup."[27]

Indeed, with the appearance of *erectus* we can see a number of attributes that we now think of as standard human characteristics – features that were a confluence of physical prowess and slowly growing intelligence, and which would have profound impact on the species and its descendants, not least *Homo sapiens*. Some of these characteristics have been the source of considerable discomfort through the ages.

Consider the hips of the Nariokotome boy. As we have seen, he had a particularly tight pelvis that gave him a narrow, cylindrical form and stance, with knees tucked underneath his body. This allowed him to keep his center of gravity under his body and prevented him from swinging his legs around all over the place when he walked. If *Homo erectus* had not evolved this gait, walking would have been enormously draining in terms of energy expenditure. Upright posture allowed *erectus* men and

OPPOSITE: How *Homo erectus* might have appeared 1.5 million years ago. This reconstruction suggests a look that is far more humanlike than the "deadly unknowing" stare of "a lion's blank yellow eyes" that paleontologist Alan Walker ascribes to the species.

SHAPING UP

Comparing a person's shinbone (the tibia) with his or her thighbone (femur) gives an unexpected insight into the geographic origins. The ratio between the lengths of these two bones is known as the crural index, and it is related to the climate in which a population lives. People who come from hot regions – such as sub-Saharan Africa – tend to be very tall and thin, like cylinders. This shape enables them to radiate heat from the body with maximum efficiency. Their crural index has been found to be high.

By contrast, people from cooler regions – such as the Inuits from the Arctic – tend to have a low crural index. Their thighbones are relatively short compared with their shinbones because their bodies tend to be more spherical and, therefore, better able to retain heat, an evolutionary adaptation to the cold of their homelands.

Importantly, when scientists began to study and measure the crural index of the Neanderthals, they discovered that this analysis gave them a low reading, which indicated that the species must have evolved in a fairly chilling environment. By contrast, when the crural index of skeletons of the first modern human beings, such as the Cro-Magnons, have been analyzed, it has been found to generate a very high reading. This suggests that our immediate ancestors came from a hot climate – a discovery that has provided scientists with one of the most important clues about the origins of modern human beings, as we shall see later in the book.

NEANDERTHAL

Neanderthals had a highly distinctive skeleton. Apart from their pronounced glowering browridges, Neanderthal men and women tend to have squat, very powerful physiques, partly in response to the rigors of their lifestyle, but also as an adaptation to the cold of their environment.

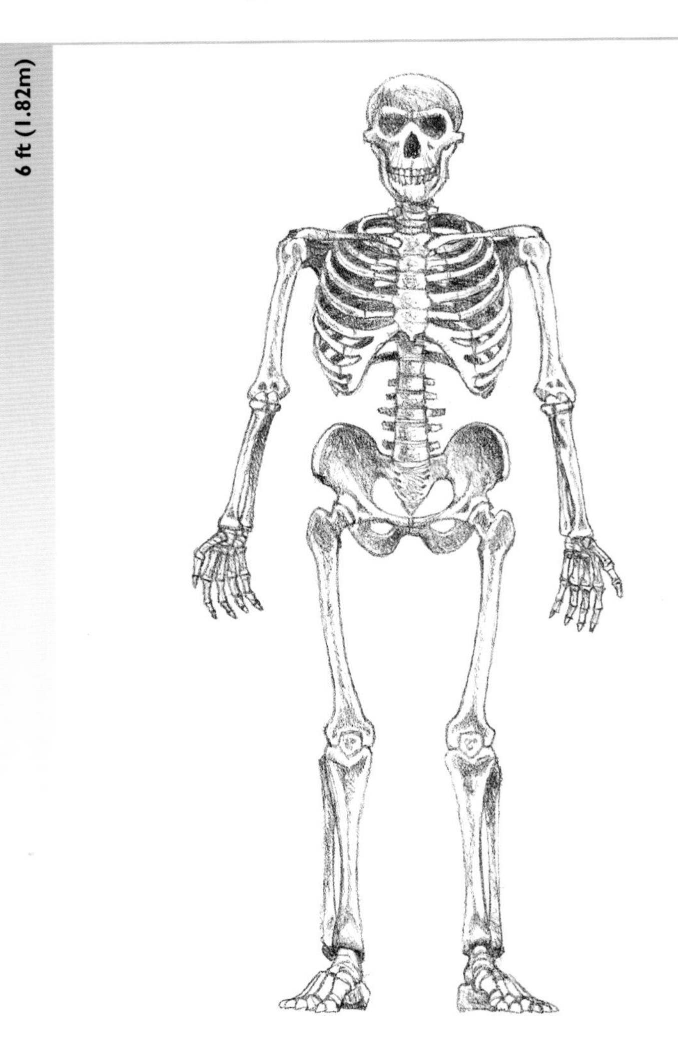

ESKIMO

Inuit eskimos from the Arctic have body shapes similar to those of Neanderthals. Their anatomies are spherical and compact, perfect for retaining heat in freezing conditions.

HOMO ERECTUS

Homo erectus, which evolved in hot equatorial Africa, had a tall, thin physique as typified by the Nariokotome boy. The species was ideally adapted to radiating heat away from their bodies.

MASAI

The Masai people of modern Kenya are also leggy, cylindrical, and tall, and they too come from an extremely hot climate. Their physique is strikingly similar to that of the Nariokotome boy.

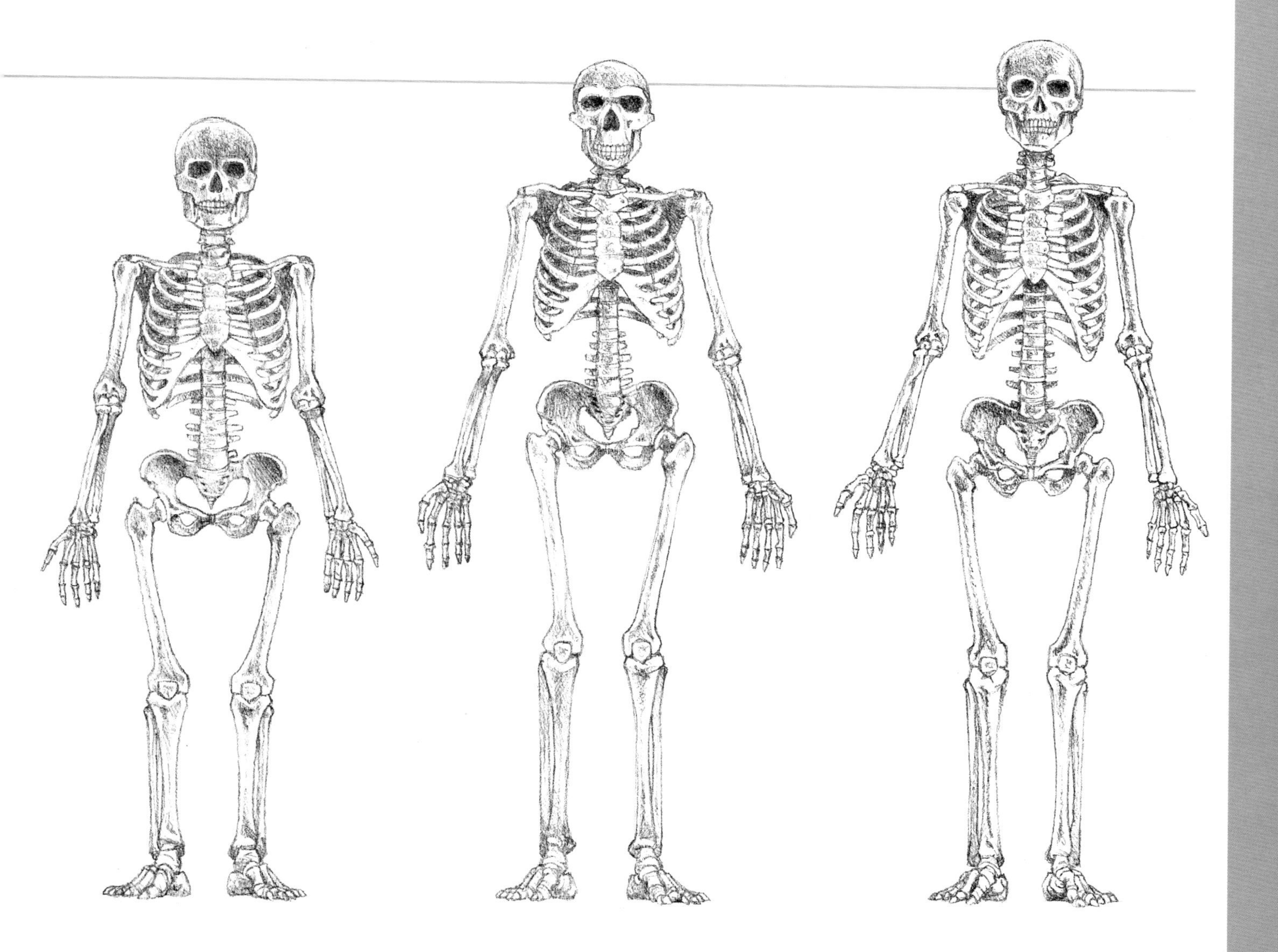

In contrast to a chimpanzee (top) or an australopithecine (middle), a modern human baby (bottom) is relatively large compared with the pelvis of its mother, and can only pass through the birth canal with great difficulty, a process that has been likened to the twisting of a cork from a bottle. A primate baby passes straight through its mother's pelvis, however, and its birth is relatively easy and painless. The birth of a human baby is far more risky and traumatic.

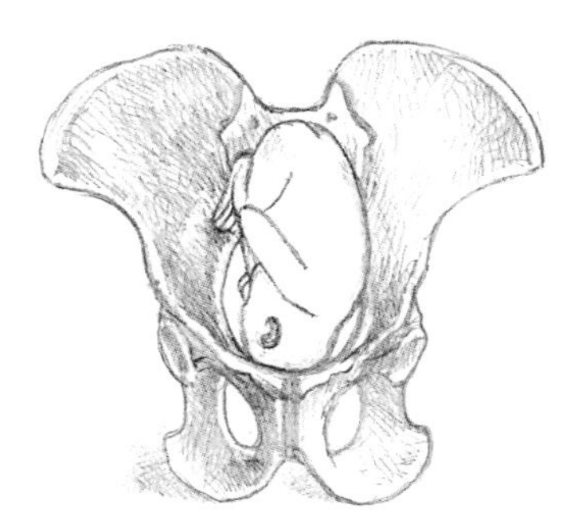

Chimpanzee

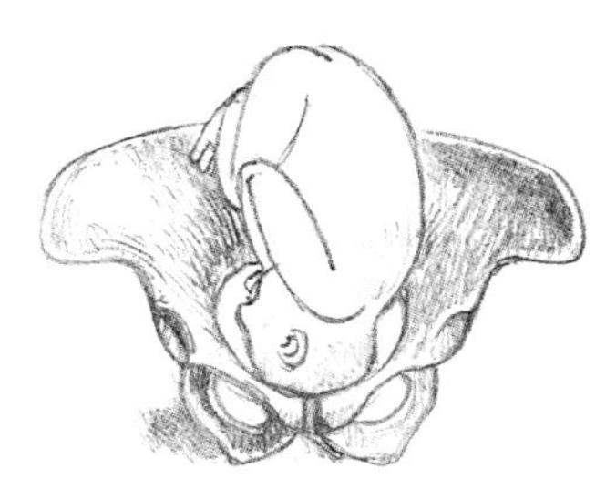

Australopithecines

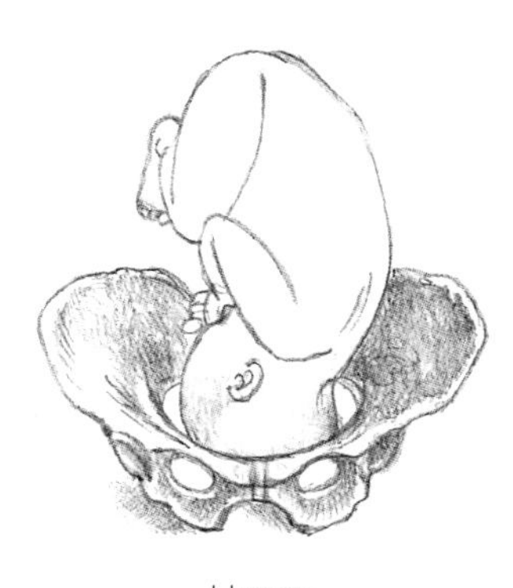

Human

women to stride across the savanna – and ultimately the globe – but at a cost: for a child must pass through the pelvis when it is born, and a tight pelvis leads to problems at birth.

"No rational animal would knowingly evolve bipedalism and then go on to develop large brains," says Aiello. "The consequences for women have been horrendous."[28] While chimpanzees go through their most crucial neurologic developments in the womb, a human is born having undergone less than half of this critical increase – because a baby's head would be too vast to push through the pelvis of any woman who had aspirations to remain a creature with an upright gait. As a result, humans evolved so that a child has to spend the first year of its life in a particularly helpless form as it completes the brain development that should have occurred while it was a fetus. This, in turn, constrains the mother. She is tied to an utterly dependent young child for a long period, yet she needs good nutrition to provide milk rich in protein, fats, and carbohydrate for her offspring. She is therefore particularly reliant on the support of her spouse and the rest of her group.[29]

This process of giving birth to immature infants almost certainly began in *Homo erectus*. And as brains evolved to become larger, the infants became increasingly immature, a trend that has continued through the ages. It was to have a mighty impact on human society, with men evolving to play greater roles as providers of nutrition and protection for their mates and children, while tribes evolved greater social cohesiveness. As a result, even today birth in humans can cause problems. Other primate babies can pass straight through the pelvis.

Richard Leakey scraping the soil from specimen KNM-ER 3733, the *Homo erectus* skull shown on page 78.

A human child has to twist through the narrowest of gaps like a cork being pulled from a wine bottle – a maneuver that requires much effort from the mother and the attention of midwives, and which remains a strikingly risky business. As a result, women tend to have slightly wider hips than men, which minimizes the trauma of childbirth, but means that the female of our species is a slightly less efficient bipedalist than the male, and is – as a consequence – a somewhat poorer walker, runner, and jumper, as is revealed in athletic records.[30]

In *Homo erectus*, we can therefore see the emergence of several key defining traits of our species. As the climate became more variable, so did our repertoire of behaviors, the most critical being the increased consumption of meat, which was obtained either by scavenging or hunting (a contentious issue that we shall examine in the next chapter). At the same time, our ancestors evolved athletic physiques, which resulted in the evolution of smooth skin and an assured bipedal gait. This latter characteristic, in parallel with the increase in brain size, meant lengthier childhoods, which in turn required men and women to form stronger social bonds. It is a convincing-sounding package that helps explain the origins of a great many facets of what it means to be human – in particular our intense gregariousness and sense of community and mutual support. But is there any direct evidence that this last attribute had its origins in *Homo erectus* so long ago?

Well, yes, there is some data – from a fossil known as the KNM-ER 1808 specimen. (KNM-ER 1808 stands for Kenya National Museums – East Rudolf 1808. The first part of this label refers to the institution where the fossil is cataloged and housed, and the middle to the region of Kenya in which it was found. Many such fossils have been discovered there, of course, so a number is given to identify it precisely, and this forms the last part of the tag. The Nariokotome boy is more properly known by scientists as KNM-WT 1500, the WT standing for West Turkana. This is the label that is printed on the boxes of his bones and that distinguishes his remains from all of the other deposits in the Vault of the Kenya National Museum.) Specimen 1808 was an adult female *Homo erectus*, spotted in 1973 by the ever-alert Kamoya Kimeu during one of his desert walks. Unlike the Nariokotome boy, this was only a partial skeleton and its bones belonged to a female who had clearly suffered from some disease prior to her death. "On the leg bones were intriguing signs of some kind of disease," says Richard Leakey. "The shafts appeared patchily encrusted with new bone."[31]

Pathologists studied her remains and concluded that 1808 had suffered from an affliction known as hypervitaminosis A or, to put it more plainly, vitamin A poisoning. This is a disease of carnivores, which is in itself a significant observation. It usually occurs when a person eats too much liver, an organ that acts as a filter for vitamin A. Most recent cases have concerned polar explorers who, in extremis, have eaten too much polar bear liver or husky dog liver and paid the consequences: hair falling out in handfuls, aching joints, loosening teeth, bleeding gums, and blood clots that eventually turn into lumps of bone. And it was these last accumulations that could be seen – in profusion – on 1808's legs, and which have led paleontologists to surmise that she must have been immobilized for weeks, perhaps months, before her death. Yet she lived – and that was because someone else took care of her, says Walker in *The Wisdom of the Bones*. "Alone, unable to move, delirious, in pain, 1808 wouldn't have lasted two days in the African bush, much less the length of time her skeleton told us she had lived. Someone else brought her water, and probably food. And someone else protected her from hyenas, lions, and jackals on the prowl for a tasty morsel that could not run away. Her bones are poignant testimony to the beginnings of sociality, of strong ties among individuals that came to exceed the bonding and friendship we see among baboons or chimps or other nonhuman primates."[32] Such an idea does, of course, depend on the accuracy of the diagnosis of 1808's condition. Her remains are about 1.5 million years old, and her bones have suffered much geologic insult in the intervening millennia. A precise identification of the cause of her death, and the length of her illness, cannot therefore be carried out with total confidence. However, the disease, whatever its exact identity, was clearly in an advanced stage – given the amounts of excess bone that had been deposited – so the woman must have been in a dependent condition for some time. It makes an intriguing story.

Whatever else, *erectus* now clearly possessed a physique honed in the hominid school of hard knocks: searing heat, ferocious carnivore competitors, and a wildly fluctuating climate. The end result was an exceptional species that had evolved in exceptional times, and central Africa could no longer contain it. Other worlds beckoned, so *Homo erectus* set off – armed with stone tools, a questionable intellect, the power to kill, and the capacity to eat meat, this last attribute providing the species with a surprise secret weapon – as we shall see in the next chapter.

CHAPTER FOUR

NEW WORLDS TO CONQUER

Eugene Dubois was an unlikely revolutionary. Born in 1858, the eldest son of a devout Dutch Catholic family, his background was formal and conservative, though, as a boy, he did reveal an unexpected fondness for collecting fossils. Nevertheless, a career as a doctor beckoned, and Dubois entered medical school in 1877. These were unsettled times, however, an era when Darwin's theory of natural selection was first giving the study of biology a new impetus and direction. Despite his religious background, Dubois was aroused by the writings of the German biologist Ernst Haeckel and became an intractable convert to the ideas of evolution – an inspiration that would turn Dubois into one of science's great romantic iconoclasts. Although he had a chance of securing a professorship in anatomy in Holland, he quit medical school – leaving behind only the enigmatic promise that he would return one day "with the missing link" – and took a job as a military official in the Dutch East Indies.[1] In 1887, he set off for the Far East to start his search for human fossils. It was from here, after all, that explorers had recently returned with evidence of the existence of some fantastically humanlike creatures – including orangutans and gibbons. These seemed eminently suitable candidates to be mankind's biological cousins and made Holland's Indonesian colonies look a promising prospect to the fledgling fossil-hunter.

Nineteenth-century fossil hunter Eugene Dubois. His discovery in Java in the 1890s of the first *Homo erectus* remains ranks as one of science's greatest tales of perseverance and endurance.

Dubois was not an easy man to get along with – at least when dealing with fellow academics. He was often stubborn and hot-headed. But he could also be extremely charming – as can be judged from the fact that he managed to persuade his superior officers that a full paleontological survey of the Dutch East Indies would be a valuable undertaking. And so, in March 1890, he traveled to the island of Java with a couple of loyal corporals and a group of prisoners who were to carry out the spade-work for his endeavors. For 18 months Dubois languished in the sweltering heat, while his convict crew toiled – until one day, in October 1891 at Trinil, on the banks of the River Solo, they found a peculiarly flattened skullcap with a prominent ridge of bone – a "brow ridge" – above its (missing) eye sockets. A year later, they dug up a thighbone, or femur, which was certainly fully human. Dubois had found his missing link, as he had promised, a discovery that "ranks among the best in the annals of scientific perseverance and perspicacity," as Stephen Jay Gould puts it.[2] Dubois named the species *Pithecanthropus erectus* (erect apeman).

The discovery was of critical importance. Dubois had found the first fossil of a genuine ancestor of *Homo sapiens*, and in making his discovery in Java, he had shown that *Pithecanthropus erectus* was probably the world's first intercontinental traveler, although Dubois did not appreciate the significance of this point at the time. He simply assumed *erectus* had evolved in the Far East, an idea that has, ironically, recently reemerged with considerable controversy, as we shall see later in this chapter. Convinced he had found the "missing link" between ape and man, Dubois returned to Europe in 1895, expecting to be hailed as a scientific hero. And although he was fêted with official scientific honors in recognition of his efforts, he was also ridiculed by many scientists for suggesting that his fossils belonged to a human ancestor. They were simply those of a giant gibbon, said his enemies. After furiously defending himself, Dubois eventually retired from academic life, still sure he had found a

ABOVE: The femur and skullcap of *Homo erectus* that Dubois found in Java. His critics claimed the bones were merely those of a giant gibbon. We now know that Dubois had unearthed the first fossil of a genuine ancestor of *Homo sapiens*.
BELOW: A photo taken by Dubois himself of the Java excavation site where these fossils were found.

OPPOSITE: Java, Indonesia. Early *Homo erectus* finds, such as those by Dubois, initially led scientists to believe that human ancestors evolved here. BELOW: Dubois' reconstruction of Java man – *Homo erectus* – which he based on the two bones that he had discovered on the banks of the Solo River. The statue now resides in the Natural History Museum, Leiden.

primate halfway-house between humans and apes.[3] He was wrong in believing his discovery was apelike, but correct in claiming that he had found key human fossils.

Today we give a slightly different name to *Pithecanthropus erectus*. We call it *Homo erectus*, while the Dubois fossil is still often given the title Java Man. Subsequent discoveries of *erectus* fossils were later made in China and at various African sites, such as Koobi Fora and Olduvai Gorge. Finally, there was Leakey and Walker's great breakthrough at Nariokotome in 1984. The geographic diversity of these finds indicates the species had become, if not ubiquitous, then certainly modestly well dispersed throughout the Old World. As Gould puts it: "Dubois' beginning, a skullcap and femur from Java, has blossomed into a well-documented ancestor, widely spread over three continents."[4]

But how could *erectus* people have traveled the lengths of Africa and Asia with only the most basic of stone technologies? And when exactly did the species begin its Old World diaspora? Both questions are important and raise issues that still give anthropologists major headaches. Indeed, far from becoming clearer in recent years, our ideas about *Homo erectus*'s global movements have grown more confused and less certain, as we shall see. However, one point is clear: in evolving the carnivorous habit, *Homo erectus* had acquired a key behavioral advantage that almost certainly helped it to take to the road, as Alan Walker makes clear. "As an herbivore, you couldn't stray far from the plants with which you had become familiar. Your range was limited. But when our ancestors turned to meat, something changed. We could take advantage of other animals' adaptations to the plants in their localities – by eating them instead. Thus, we could spread over enormous distances. We conquered the world because we could eat meat."[5] Nevertheless, the world's climates remained perilously uncertain at this time, and without the sophisticated social behavior and complex tools of their descendants, *erectus* people must have found their peregrinations a daunting process. However, they succeeded – as we can ascertain from the *erectus* fossils found in such wide-ranging places as Algeria, South Africa, China, Georgia, and Java – including a particularly important fossil of a child's skull discovered in 1936 near the town of Modjokerto in Java, close to where Dubois made his great discovery. Most are dated as being between 1.8 million and 800,000 years old.

From this data, a standard picture of *erectus*'s movements began to emerge. About 2 million years ago, somewhere in central, eastern, sub-Saharan Africa, the species emerged as the robust, modest-brained creature that was subsequently to

be typified by the Nariokotome boy. Then, according to the theory, around a million years ago our ancestors began traveling. Mile by mile they slowly extended the range of their hunting and gathering. They passed into the Levant, before moving east into Asia, reaching Java about a million years ago.

This timetable of *erectus*'s movements was based on the age of the fossils found in the Far East. Unfortunately, some of the data lacked reliability, as Roger Lewin of Harvard University's Peabody Museum points out: "One problem with fossil-collecting in Java was that it was often performed by local farmers, who came across specimens in their work or developed a talent for finding them. The issue of provenance of a fossil, or its exact location in its sediments, was therefore often a serious problem. Accurate provenance is essential if a fossil is to be reliably dated. This caveat applies particularly to the Modjokerto skull found in 1936."[6]

It was a problem that stretched right back to the days of Dubois, who did no sweltering excavating himself, unlike his successors at Nariokotome. Instead, he sat on his veranda, miles from the digs, and his staff brought their finds to him. The exact positions of his two great discoveries – the Trinil skullcap and thighbone – were never known, so the dates of those fossils could not be determined.[7] Their antiquity was assumed to be similar to those of subsequent discoveries in the region, although few of these were dated with any real confidence. Java, unlike the Rift Valley, lacks the definitive geologic history that makes dating a fairly exact science in Africa. As a result, scientists have often had to assess the antiquity of Javanese hominids by looking for other animal fossils nearby. For example, if bones from an extinct species of rodent are found, and it is known that this animal disappeared from the fossil record 1 million years ago, then it can be assumed that the sediment containing the bones is more than 1 million years old. It is neat, but not foolproof, for you have to know exactly where you found your rodent and be certain that it was in the same layer of soil as your human fossil. This was always a problem in Java.

Then, in 1992, two German researchers announced that they had found a *Homo erectus* jawbone at Dmanisi in Georgia, east Asia. They argued that it was between 1.8 and 1.6 million years old, basing their assessment on other animal bones found in the same sediments. Subsequent research has indicated that these dates may be exaggerations. Nevertheless, the timetable of *erectus*'s travels was beginning to look a little dubious, failings that were confirmed two years later with the dropping of a true paleontological bombshell by two of the world's most distinguished experts on fossil-dating: Carl Swisher of the Berkeley Geochronology Center in California and his colleague Garniss Curtis, the scientist who was responsible for dating the Laetoli footprints, which we came across in Chapter One. The pair decided to reexamine the volcanic deposits in which the Modjokerto skull had been uncovered. The skull had originally been dated as being around 1 million years old by a Japanese-Indonesian team who used a dating technique called fission-tracking (see pages 100-101). Fission-tracking reveals the age of particles of volcanic rock buried along with the fossil. Unfortunately, it is not very precise. An alternative technique – known as potassium-argon dating – is better, but this was ruled out because the sediments did not contain enough potassium. The answer, Swisher and Curtis decided, was to exploit a newly developed version of potassium-argon dating called single-crystal laser fusion. As the name implies, this technique requires only a tiny amount of material – a single crystal of volcanic rock was sufficient.

Scientist Carl Swisher who, along with his colleague Garniss Curtis, used newly developed dating methods to ascertain the true age of the *Homo erectus* fossils found in Java. The results astounded scientists and showed that *erectus* made its 10,000-mile (16,000-km) journey from Africa to colonize the Old World 1.8 million years ago.

So the pair collected a few of these volcanic crystals from Modjokerto and took them back to their laboratory in Berkeley, where they produced startling results: the skull was 1.8 million years old – almost twice the previous estimate, revealing that *erectus* arrived in Java at least 800,000 years earlier than was previously thought. In addition, Swisher and Curtis redated an *erectus* fossil from Sangiran, Java, as being 1.6 million years old. Up till that point, the oldest known *Homo erectus* specimen was the KNM-ER 3733 fossil from Koobi Fora in Kenya, which was thought to be about 1.8 million years old. Now a similar date was being given to a specimen that had been found 10,000 miles (16,000 km) from the species' birthplace. The news caused a scientific sensation. "This is just overwhelming. No one expected such an age," the anthropologist F. Clark Howel of the University of California told *Time* when the magazine devoted a cover story to Swisher and Curtis's work in March 1994.[8]

If correct, *erectus* must have fairly leapt out of its cradle and left home for Java with almost indecent haste. After emerging in central Africa 2 million years ago, it appears to have set off almost as soon as it could stride the savanna in order to get to

DATING THE PAST

ABOVE: The potassium-argon dating method, developed at the University of California, Berkeley, is used to determine the absolute age of the volcanic rocks and deposits in which paleontologists have found hominid fossils. This technique was first used in the 1950s on sediments taken from Olduvai Gorge, Tanzania.

A battery of different procedures is employed by scientists in order to discover the antiquity of ancient bones or prehistoric tools, and most involve the measurement of rates of radioactive decay. A good example is provided by the best known of all these techniques: radiocarbon dating.

The carbon that forms the bodies of all living objects is made up of two isotopes – carbon-12 and carbon-14 – which are chemically indistinguishable from each other. However, the rarer carbon-14 is slightly radioactive and slowly decays over the millennia into nitrogen. This means that if a scientist finds an old bone or a wooden implement, he or she can measure its carbon-14 content to reveal its age. If it contains very little carbon-14, this shows that the object is of considerable antiquity, while if it possesses quite a lot, this reveals that it is relatively modern. Developed by American scientists in the 1940s and first used to date ancient Egyptian artifacts, this form of dating suffers one key limitation: carbon-14 decays relatively quickly. That means that in objects older than about 40,000 years, there is simply not enough of the isotope left to permit any kind of accurate dating.

To get around this problem, scientists have exploited the radioactive decay of a different element: potassium. About 0.01 percent of all potassium is made up of a radioactive isotope called potassium-40. Since it decays very slowly into an isotope of argon called argon-40 (the process takes millions and millions of years), by measuring how much argon-40 gas a rock contains, scientists can measure its age. Potassium-argon dating was first developed in the 1950s and played a crucial role in determining the ages of the sediments surrounding the fossils that were then being found in Olduvai Gorge and other parts of East Africa.

Modern versions of the technique – such as single-crystal fusion dating – allow scientists to find the age of an object from a tiny fragment of material. In addition, other technologies are expanding paleontologists' ability to date specimens. These include electron spin resonance, thermoluminescence, optically stimulated luminescence, and uranium series dating.

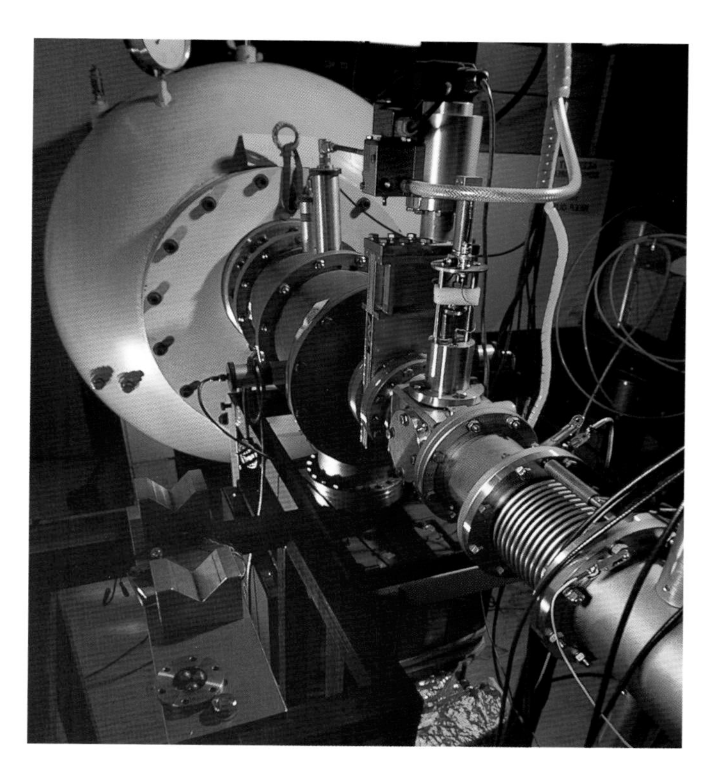

RIGHT: The device used for counting carbon-14 atoms in a fossil or other organic object sample in order to determine its age. BELOW: The various techniques used to date fossils and the age ranges that they cover.

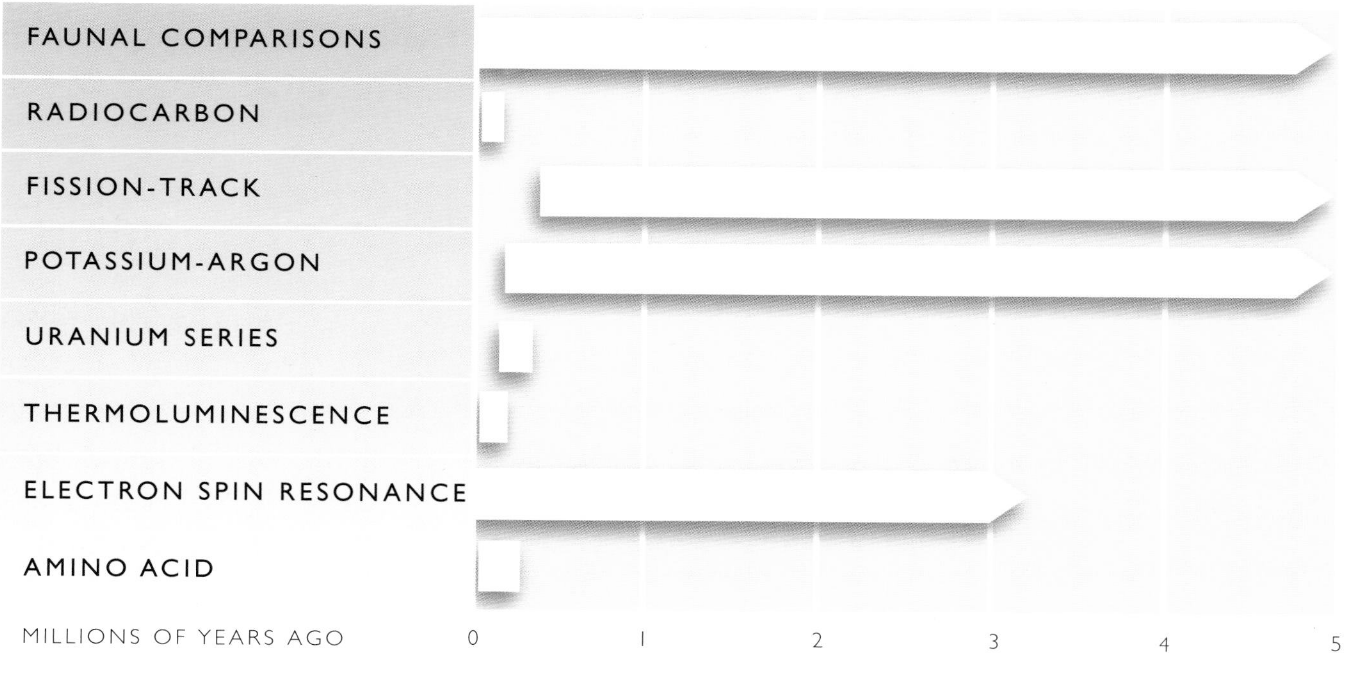

the Far East by 1.8 million years before present. This is not an impossible concept. Even at a relatively sedate pace of population expansion of 10 miles (16 km) per generation, *Homo erectus* could have got from East Africa to East Asia in a mere 25,000 years. As Swisher says, "Elephants left Africa several times during their history. They didn't need stone tools to get to Asia. And that goes for lots of other species. It is just because we are dealing with humans that people have difficulty dealing with a migration of such rapidity. We somehow think human behavior should be different or special, but we shouldn't let that cloud our thinking."[9]

Not every scientist is comfortable with such a rapid migration, and some argue that the facts best fit a very different theory: that *erectus* actually evolved somewhere outside Africa. From this as yet unknown land they later migrated back to Africa and on to Java. The hypothesis assumes that *erectus*'s predecessors, perhaps *Homo habilis*, were really the first intercontinental travelers, and that a key part of our evolution – the appearance of *erectus* from *habilis* precursors – took place outside Africa, somewhere in Asia. Such ideas can at least reconcile the dating problem that Swisher and Curtis have thrown up, but they do leave us with a rather messy alternative in which our ancestors seem to have wandered all over the world as they evolved. "It is a fascinating idea, but one undermined by the lack of any plausible ancestors to *erectus* in Java or anywhere in Southeast Asia," Walker says. "There is not a scrap of a hominid until *erectus* arrives. In contrast, in Africa for 2 million years before *erectus* appears, there are thousands of specimens of several species of undoubted hominid that might be ancestral to *erectus*. The immediate predecessor to *erectus*, the troublesome *Homo habilis*, makes a fairly likely ancestor. It would take a huge number of new fossil discoveries in Southeast Asia to overturn this pattern."[10] Swisher agrees: "It is possible that *erectus* could have evolved in Asia but, until we find a *Homo habilis* fossil in Asia, I can see no reason to believe such an idea."[11] Nevertheless, Eugene Dubois would have been delighted to find that the Javanese homeland that he had postulated for early mankind was now reemerging as an important new arena for human evolution.

In any case, some paleontologists are still reluctant to accept Swisher and Curtis's new dates. Perhaps volcanic rock fragments older than the skull had eroded from the slopes of nearby volcanoes and been buried. Or perhaps the bones had been washed down into older sediments. And then there is the problem of the Modjokerto skull's exact resting place. As we have seen, most Javanese skulls were found by farmers or local collectors. In this case the find was made by a local man, called Tjokrohandojo.

In 1975, he explained to an Indonesian paleontologist, Teuku Jacob, where he had found the skull in 1936, and Jacob in turn told Swisher – almost 20 years later in 1994.[12] This reliance on human recollection, over many decades, raises obvious worries about whether the deposits that Swisher and Curtis tested were the right ones or not.

But the Modjokerto fossil was not the only one investigated by Swisher and Curtis. They had also looked at a site at Sangiran and had found an age of equal antiquity for the *erectus* fossils that had been uncovered there. As Swisher points out, "it is highly unlikely that you would get the same kind of errors in both places."[13] In addition, there were the tests they carried out on the Modjokerto fossil itself. The underside of the skull had been painted black many years ago by an unknown culprit, and Swisher and Curtis spotted an odd lump. "We noticed there was an odd bump in this painted area that we knew wasn't part of the anatomy," says Swisher. "So we scraped a little of it away and realized that the underside of the skull was coated with pumice."[14] They were allowed to remove one tiny piece of this volcanic rock, which they tested in their laboratory and found that it had a similar chemical "fingerprint" to minerals from the sediment layers they had assumed were the resting place of the Modjokerto skull. As far as he and Curtis are concerned, the inescapable conclusion to be drawn from their research is that *Homo erectus* left Africa almost a million years earlier than scientists had previously thought. And although this scenario may seem strange, it would explain one *erectus* oddity: the differing tools it used in Africa and Asia.

As we have seen, *erectus* probably evolved about 2 million years ago. Then, about 500,000 years later, it left evidence in Africa that it had developed a new type of tool kit: the Acheulian assemblage (which gets its name from St Acheul in France, where these implements were first identified in the 1830s). For the first time, relatively large stone implements were clearly being made: cleavers, picks, and handaxes. The last – a teardrop-shaped tool sometimes called a biface because it was carefully fashioned on both sides – is typical of this equipment. Compared with the crudely shaped volcanic pebbles and the small rock-flakes of the older Oldowan tools, Acheulian implements were clearly made to a template in their makers' minds. They were constructed with skill and strength, and included a range of specialized tools that indicated their creators were capable of some degree of forethought. As for their precise mode of use, a few scientists believe *erectus* may have thrown the handaxes to bring down deer or other animals, although most believe they were predominantly employed as heavy-duty knives. This idea is backed by Indiana University archaeologist Nick Toth, whose experiments

have shown that a handaxe could slice the toughest type of hide, including an elephant's, while microscopic analysis of the blades has revealed traces of meat, bone, and wood.[15]

Given the handaxe's usefulness, one would expect to find it everywhere that we find signs of *Homo erectus*. But that is not the case. Scientists have uncovered an odd discrepancy: although Acheulian handaxes appear to have been used widely in Africa and western Asia, they are almost entirely absent from the Far East. For some reason, it appears that the Stone Age equivalent of the Swiss Army knife was never used by *erectus* people in eastern Asia, despite its obvious usefulness in other parts of the Old World. But why not? One solution is put forward by Geoffrey Pope at Illinois University. As he points out, the Far East is rich in bamboo, an extremely versatile material. It could have replaced stone as the basic working medium of tool-makers. Not a trace would have survived the intervening eons.

It is an intriguing, but ultimately unprovable, hypothesis. In any case, the new dating by Swisher and Curtis offers a far more enticing explanation. If *erectus* people left Africa so long ago – between 1.8 and 2 million years – their emergence would have predated the invention of Acheulian technology. In other words, they walked into Asia with an out-of-date kit. It was left to the men and women who stayed at home to come up with an invention that was to be one of the most stunningly durable innovations in human history. As we have seen, the Acheulian handaxe was still in use a million years later – but not in the Far East. The tools found beside *erectus* fossils in this part of the world are much cruder and more like

OPPOSITE: Acheulian handaxes and cleavers were first made around 1.5 million years ago by *Homo erectus*. Although widely used in Africa, Acheulian technology never evolved in the Far East.

ABOVE: Casts of the skull, jaw, and teeth of Peking Man (a member of the *Homo erectus* species) found in the Zhoukoudian caves, 25 miles southwest of Beijing, China.

CHILDHOOD'S END

Africa provided the fertile ground for the genesis of humanity – although our ancestors were not tardy about shrugging off their birthright. As soon as they had reached their evolutionary adolescence, groups of our ancestors quit the continent and headed for new lands with almost indecent haste. The main driving force behind this diaspora was that early humans were becoming more and more efficient omnivores. No longer tied to ranges in which plants grew, *Homo erectus* was now capable of scavenging, and possibly hunting, for meat – a gift that triggered its first migrations from Africa.

Scientists originally thought these movements began about 1 million years ago. However, recent dating work in Asia suggests that our African exodus occurred much earlier, possibly closer to 2 million years before present – not long after *Homo erectus* first evolved. Further confirmation of this startling idea was provided in July 1999 when a joint German-Georgian team of scientists uncovered two perfectly preserved humanlike skulls at Dmanisi in Georgia, on the southern boundary of Europe and Asia. Crucially, these fossils were dated as being between 1.6 and 1.8 million years old. It seems mankind had not only emerged out of Africa very early in its evolution and moved eastward, it had also migrated northward toward the less balmy lands of Europe.

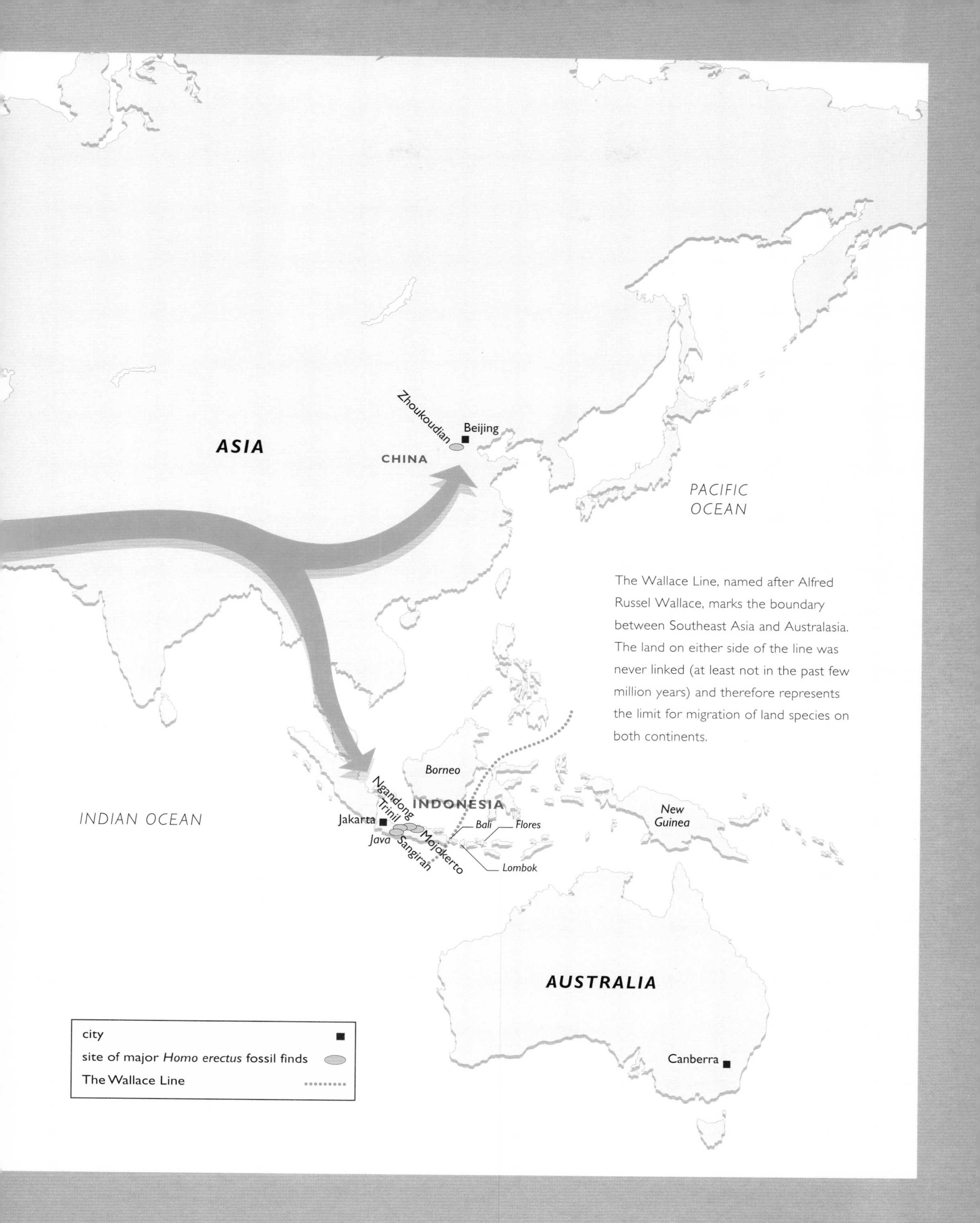

The Wallace Line, named after Alfred Russel Wallace, marks the boundary between Southeast Asia and Australasia. The land on either side of the line was never linked (at least not in the past few million years) and therefore represents the limit for migration of land species on both continents.

Dragons' teeth: the remains of fossils of various animal species including two teeth from a hominid (top right). The Chinese thought ancient bones and teeth could cure illnesses, and ground them up to make medicines. Some of the first fossils found in China were bought from drugstores, as were these specimens.

those Acheulian predecessors, the Oldowan tool kit. That is what early *erectus* emigrants took with them when they left Africa, and that is what they stuck with over the eons. When later tribes of *Homo erectus*, bearing the new technology, tried to follow them eastward, they may have been repulsed by existing populations of the species, or perhaps they never reached them because the latter had become geographically isolated. The suggestion is intriguing but is also difficult to prove.

Part of the problem, as usual in paleontology, is lack of evidence. In the case of *Homo erectus*, this absence is particularly galling, given the fate of some of the fossils. One of the first *erectus* specimens to turn up after Dubois' initial discovery was a skullcap that was discovered inlaid in layers of limestone in a cave near the village of Zhoukoudian, near Beijing. It was a distinctly fortuitous discovery, even by paleontology's normally serendipitous standards, because the hills in this region had been mined for hundreds of years for their fossils: "dragon bones," which were supposed to have fantastic healing powers. The bones and teeth of extinct animals and hominids were ground up and sold as medicine. As one author once put it, we will never know how many powdered bones of early humans "have passed harmlessly through the alimentary canals of dyspeptic Chinese."[16]

Our discussion of *Homo erectus*'s tools raises another critical question: were they the tools of a hunter or a scavenger? We know that, in Africa at least, stone weapons were involved in obtaining meat. But was this ancient butchery carried out by hunters who brought down their game? Or was it the work of crafty opportunists who used weapons to chase carnivores from their kills or to finish off sick or wounded animals? In short, was it *Homo erectus*: hunter, or scavenger? In the 1960s and 1970s, most anthropologists would have supported the former option; man the hunter was a popular notion in those days. Since then, scientists have revised their ideas and, from studies of major sites, particularly those in Kenya's Olduvai Gorge, have concluded that scavenging must have played an important role in obtaining meat. Ancient humans appear to have transported stones to key sites and then knapped them into tools. They probably also took bones with meat on them to these sites, the result of scavenging (and possibly of some hunting), says Roger Lewin. "Lacking weapons to kill at a distance, as humans did until late in prehistory, hunters could achieve only very limited goals and might not have qualified as hunters in the commonly understood sense. Scavenging, on the other hand, would have been both technologically and ecologically feasible."[17] In short, *erectus* was both hunter and scavenger, as anthropologist Jonathan Kingdon points out. These early humans were probably quite systematic about laying ambushes. "They are also likely to have been skilled at relieving the smaller or more solitary carnivores of their prey."[18]

No matter what the exact division was between those two methods of obtaining meat, it is clear that it had a dramatic effect on our ancestors. Meat allowed us to free ourselves from the dietary shackles of our African homeland. It did more than that, however: it made us brainy. Easy to digest and rich in energy, meat provided the vital resources that our expanding brains demanded. Our powerful digestive systems, previously needed to extract the meager nutrients available in vegetation, were freed of some of the rigors of their work. The new diet provided mothers with high-quality food for the brains of their developing babies, and provided continuing neurologic sustenance as those infants grew up. And not just meat, but fat and bone marrow – easily digested, energy-rich foods that

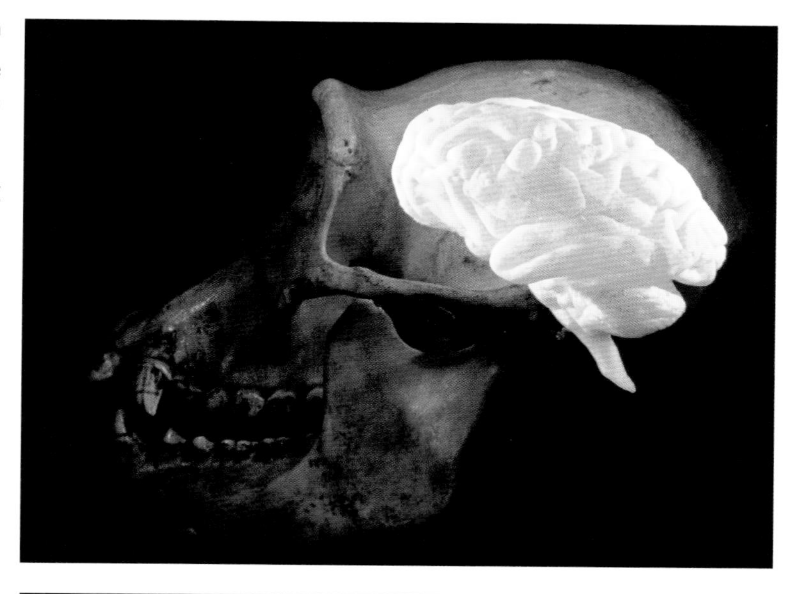

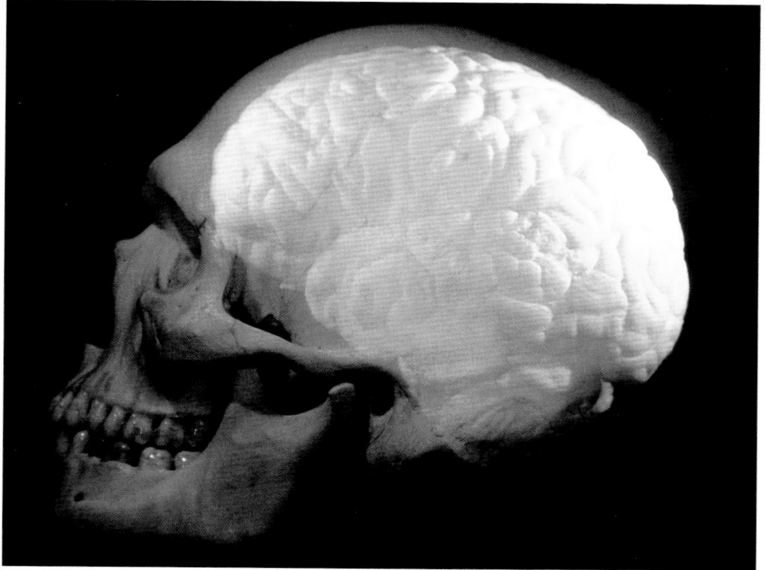

When *Homo erectus's* diet switched from a largely herbivorous one, such as that eaten by chimpanzees, to one including meat, fat, and marrow, it provided the energy for their brains to expand, a process that continued with *Homo heidelbergensis*, Neanderthals, and finally *Homo sapiens*. Seen together, the difference in size between the brain of a chimp (top), our closest "relative," and that of a modern human is remarkable.

permitted the evolution of smaller stomachs, which in turn saved internal energy, says Leslie Aiello. "The surplus was used to feed our brains, which began to grow significantly at this time. It was a loop. We started to eat meat, got smarter, and thought of cleverer ways to obtain more meat, although learning to obtain other rich, but easily digestible foods, such as tubers, was probably also involved."[19]

Indeed, this last aspect of *erectus* meals may actually have been more important than the former. Researchers led by Harvard anthropologist Richard Wrangham argue that calorie-rich tubers – of which potatoes, turnips, cassava, yams, and manioc are modern examples – might have been the nutritional triggers that set off hominid brain growth. He points out that, at this time, the climate had begun to dry out, making fruit, nuts, and possibly animal prey much scarcer. By contrast, tubers, which still grow in profusion in countries such as Tanzania, would have been unaffected by these weather changes. In addition, it is known that other tuber-eating animals – pigs and mole rats – thrived at the time, and chimpanzees today have been observed eating them, suggesting that australopithecines may also have consumed them. However, tubers may not have become a significant part of our diet until *erectus* men and women tasted one that had been baked in a grass fire – perhaps set off by a bolt of lightning – and so learned the value of cooking. At this point, heat turned "hard-to-digest carbohydrates into sweet, easy-to-absorb calories," states Elizabeth Pennisi in *Science*.[20] So, were tubers the real stimuli for our big neurologic start-up? The theory is not without critics. Loring Brace of Michigan University says the first clear evidence that our predecessors were building hearths on which tubers could have been cooked does not appear until about 250,000 years ago. "The application of fire was a late thing," he says. "I think Wrangham is on the wrong track."[21] This remark raises the important question of humanity's mastering of fire (see page 160). As we shall see, there is evidence of charred material associated with human fossils as old as 1.4 million years, but researchers cannot agree whether these were caused by bushfires or human action. Certainly, there is no unambiguous proof that humans were igniting sticks or cooking baked potatoes at the time of *Homo erectus*. Absence of proof is not proof of its absence, of course.

Meat is not only high in protein but also easy to digest, therefore diverting energy from the guts to the brain. Marrow, extracted by pounding bones with stone tools, is especially high in nutrition.

In other words, *erectus* folk may well have been the first meat-and-potato people, although in what proportions these foods made up their diets we cannot yet say. Nevertheless, some kind of dietary revolution must have begun, for the human gut today is the only energy-demanding organ that, compared with other mammals, is markedly small in relation to body size while the brain is strikingly large. The latter should weigh about 10 oz (280 g) for a mammal of our size. In fact, the human brain today weighs 3 lb (1.4 kg). Similarly, our gut – including stomach and intestines – is about half the expected size. "And small guts are compatible only with high-quality, easy-to-digest food," adds Aiello. We can see signs of this digestive diminution in the Nariokotome boy. In apes and australopithecines the ribcage is shaped like a pyramid that gets larger as you move down the body – to make way for a large stomach and coils of intestine. *Homo erectus* was the first hominid to have a barrel-shaped ribcage that opens out to make way for the lungs and then contracts over the small gut area. At the same time, we see clear signs of brain enlargement.

Of course, eating meat and tubers does not make all omnivores clever. It was just that, in the case of early mankind, it permitted an already-smart creature to get even smarter. Until then, our brain size was limited because, as Aiello puts it: "You cannot have a big brain and big guts. Providing energy – i.e. food – for both would have kept you so busy you would have had less time and energy for reproductive behavior – which is not a good move if you want to avoid extinction."[22]

This hypothesis does not explain why humans turned to a wider diet in the first place, but it does explain its success. Africa had become more arid and desiccated, and humans – like *boisei* and *robustus* – had either to become specialized devourers of vegetation or turn into omnivores. In choosing the latter, our ancestors developed a new, flexible diet that allowed their brains to grow

Compared with a chimpanzee and an australopithecine, the gut of a modern human being is noticeably smaller, about half the size. As a result, human ribcages have become narrower.

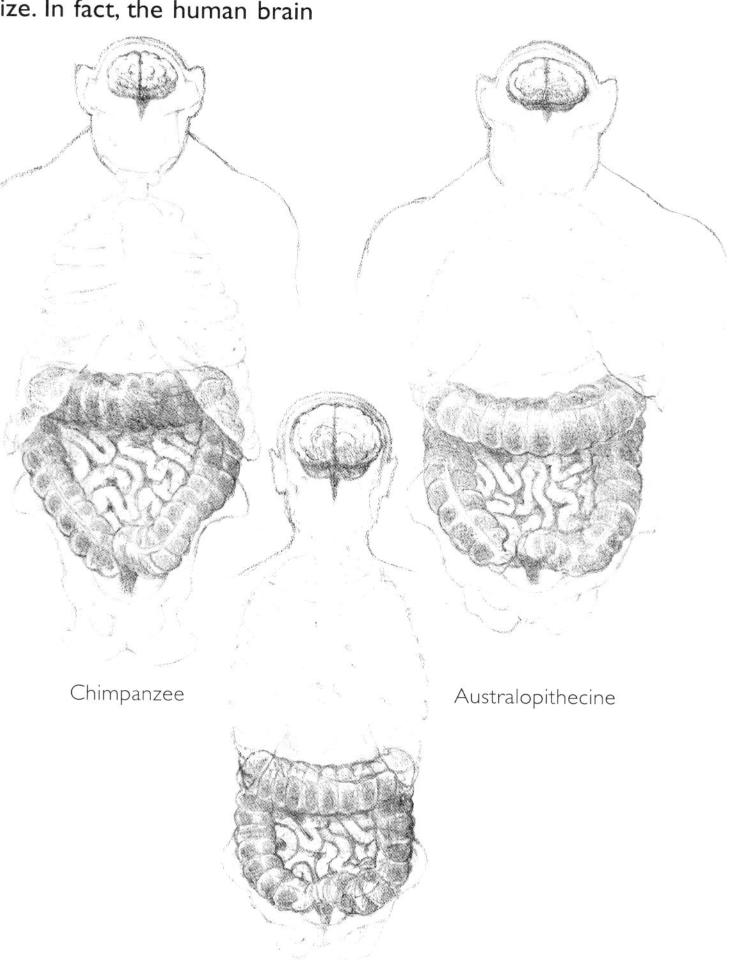

FOOD FOR THOUGHT

Big brains come with a price tag: they are highly expensive in terms of the energy they use up. Our ancestors' cranial increase would therefore have been impossible had they not turned to the consumption of foods rich in calories, such as nuts and tubers, as well as meat, which also contains a plentiful supply of fatty acids – essential building blocks for brain growth. It was this dietary expansion that provided the nutritional kick-start for humanity's intellectual growth.

Such culinary changes also stimulated a major shift in behavior. The predecessors of *Homo erectus* were the australopithecines, who were, by and large, vegetarians. Now mankind had to think about gathering a far wider range of foods, especially meat, which would have been obtained either by scavenging or by hunting. At first, the former practice almost certainly predominated. Attracted by circling vultures, our ancestors would have moved in to try to steal the leftovers of a big kill: an antelope left in a

tree by a leopard, or a large animal, such as a wildebeest, that had been brought down by lions. Scavenging from under the noses of such dangerous creatures would have been a rather risky business, but it would also have brought rich rewards. Breaking open the marrow-rich leg bone of an animal such as a wildebeest would have provided a massive dose of calories in a single, swift meal.

This calorific input would have provided the resources needed to fuel the swelling brains of humans, an input that would – in turn – have improved their intellects and therefore our ancestors' ability to find their own meat, for instance by enhancing their ability to retain complex mental plans of resources and to cooperate in dangerous hunts. Even then, humans would have concentrated on small prey. Indeed, it would be a long time – many hundreds of thousands of years – before they would develop into the highly efficient hunters that typify the later stages of our evolution.

beyond the size that had been previously imposed by their vegetarian diets. This course, by chance, freed energy and thus permitted brain growth, and that, in turn, made us more efficient omnivores. A loop was created – one that established and rewarded increased intellect, creating minds that were capable of real intellectual tasks as well as social complexity.

We can therefore see why such dietary changes were slowly adopted by *erectus*, and what effect these changes began to have on our lineage. However, there were consequences, although these had less to do with our neurons and more to do with our evolution as social animals. With increased reliance on meat and tubers for food, children would have been less able to forage for themselves, and parents would have been obliged to play a greater role in feeding their offspring. "This would have been one of the most significant events in human evolution," says Aiello.[23] "When mums began to share their own food with their children, it opened up previously unimaginable horizons. They could provide nutrition in forms that their offspring didn't have the strength, knowledge, or coordination to get for themselves. A new dietary flexibility was introduced – for all the family."

This nutritional advantage had further, profound consequences. Better fed, our ancestors began to live longer. We can see this today by correlating the body size of various mammals with their lifespans. According to this scale, humans should die around the age of 40. We live much longer because we eat healthier, more nutritious foods. And the onset of human longevity would, in turn, have had its own considerable impact on social structure. For a start, there would have been increasing numbers of older people to provide advice. They might have remembered what they did during a previous drought, for example, adding a new flexibility to our repertoire of behaviors.

But that may not have been the only advantage. One particularly intriguing idea has been suggested by Kirsten Hawkes of the University of Utah. The emergence of long-lived women may have been especially important, she suggests – for it led to the growth of "granny power."[24] Women reaching middle age would have been able to help their daughters with the tricky business of foraging for food. These daughters and grandchildren would have thrived. Young mothers would have had less work to do and could divert more energy into having offspring – thus swelling *erectus* numbers and increasing its urge to move on to new lands. "We evolved a society in which each person had two mothers, a mum and a granny," says Hawkes.[25] Indeed, grandmothers became so important that menopause evolved to stop them having children of their own late in life. Grandchildren became a better investment than children, which helps to explain one of the most puzzling features of human biology: the fact that we live long past our reproductive age.

In all, it is an intriguing suite of changes in our behavior, and we can see how successful these changes were for our ancestors. By 1 million years before present, *Homo erectus* seems to have settled in most of the Old World. It had made a refuge in Java that was to last a startlingly long time, for example, and had found homes throughout much of Africa and Asia. Our stage is therefore set for the next major drama in human evolution: the destiny of *Homo erectus*. Unlike so many other lineages that we have discussed in this book, extinction was not to be its fate. It would evolve, although exactly where and into what is the intriguing question. *Erectus*'s arrival in the Far East was certainly rapid, if the latest dating studies are correct, but its fate then becomes obscured in the mists of prehistory. However, *Homo erectus*'s migration to the west – into Europe – has left behind far more tantalizing indications of its presence, and these are the discoveries that are causing the real scientific excitement about this remarkable species, and which will be the focus of the next chapter of this book.

CHAPTER FIVE

A HEAD START

Of all the sites that have changed the way we think about our own species, few compare with "La Sima de los Huesos" – The Pit of Bones – either for the majesty of its hidden treasures or for the sheer drama of its location. The 300,000-year-old remains of 32 humans have been scraped from the sticky sediments of this vertiginous shaft, which drops 50 ft (15 m) from the deepest recesses of Atapuerca Cavern, near Burgos in northern Spain. In a field of science in which the finding of a single tooth can make headlines and reputations, such riches – complete skulls, spines, and ribcages of our primitive predecessors – are without precedent, and are now providing paleontologists with precious information about our evolution's penultimate phase: the transformation of robust hominids into intelligent humans.

When *Homo erectus* emerged from Africa, it was still a creature with only modest aspirations to intelligence. Such deficiencies were not destined to last forever, though, and were to change dramatically – but scientists were at first hard-pressed to find evidence for the transition that they knew must have occurred. Only a few scraps

Paleontologist Juan Luis Arsuaga descends into the Pit of Bones at Atapuerca, northern Spain.

of human material from this transitional period were known until the wonders of La Sima were revealed. Indeed, in the few years that the pit has been excavated, the fossils it has yielded have come close to monopolizing our vision of the period, for they now account for three-quarters of all human remains from the period between 100,000 and 1.5 million years ago. If you consider merely the bones of the hand, there is one fragment from China, another from southern France, but more than 300 from La Sima.[1]

It is a paleontological treasure trove, although in this case the booty is a little difficult to reach. Just to get to the top of La Sima, you have to scrabble nearly half a mile through twisting limestone potholes. At one point, the intrepid visitor has to squeeze along a tight, narrow tunnel with only bat guano for lubrication. After half an hour of scuffling in the dark and filth, you reach the pit: a dismal shaft the height of a four-story building. A thin wire rope ladder, with tiny steps, snakes flimsily into the void. Fortunately, a safety harness is provided, for the wire ladder can spiral awkwardly before the unwary visitor is deposited onto La Sima's clay floor.

From here you have to slither down another 30 ft (9 m) of mud, before sliding into a chamber the size of a walk-in closet, in which scientists – working on low scaffolding, flicking specks of dirt from bones protruding from red sediments, and recording on charts the exact position of each unearthed piece – have slowly uncovered the cavern's secrets. Not that the pain of being in La Sima lasts long. After a few hours' work, oxygen in the chamber runs out. If you were stupid enough to try to have a cigarette, you could not even light a match in the exhausted air. The only sensible thing to do is make a rapid exit, which is no easy trick if you have to scuttle up a wire ladder and crawl through a limestone maze.[2]

Clearly, it takes a special sort of person to work at Atapuerca. But then, it is a special place. Some of the first humans to set foot in Europe appear to have settled near here, and they left tantalizing glimpses of their presence. For example, at a site called Gran Dolina, in an old railroad cutting a few hundred yards from Atapuerca Cavern, scientists have found bones and stone tools about 800,000 years old – half a million years older than the fossils from La Sima. The people who left behind these intriguing remnants probably came to the Atapuerca region because of its cool oak forests, streams, and game. They used crude stone tools to hunt and eat horses, bison, deer – and each other. As the leader of La Sima's investigators, Professor Juan Luis Arsuaga of Universidad Complutense, Madrid, puts it: "It is quite clear from the

grooves on the earliest human bones we find here that corpses were defleshed. And the only reason for cutting off the flesh would have been to eat it."[3] However, it is unclear whether cannibalism was practiced by these colonizers in order to provide an alternative source of nutrition to bison meat or venison, or as part of some ritual – for instance, one in which the flesh of an ancestor was venerated through its consumption.

The Atapuerca region began to give up its secrets about 20 years ago. Although the new discoveries were critically important, they left many questions unanswered. Were the early settlers African in appearance or had they begun to evolve European features in response to the continent's cooler weather – large noses and squatter bodies, for example? And how big were their brains – were they beginning to approach the size of those of modern humans? Just how intelligent were these people? Answers to such queries are essential, for without them we cannot make sense of the final path that mankind followed as it haltingly moved toward its present status. Fortunately, a continent-wide series of digs has begun to shed light on this intriguing period of human prehistory, sites that will take us on a paleontological pilgrimage through Europe – to Britain, France, Germany, and Greece – before returning to Atapuerca.

So let us start our European tour at Gran Dolina in Atapuerca, leaving its

OPPOSITE: Gran Dolina, near Burgos in northern Spain, where scientists have found some of the earliest remains of human beings in Europe. One of the bones found there (below) has grooves on its surface that suggest our early ancestors may have stripped off the flesh.

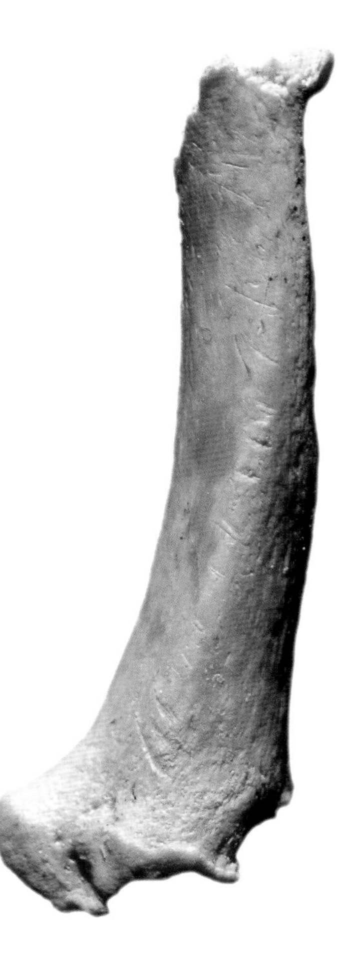

cavernous counterpart, La Sima, for the end of this chapter. Dolina was settled by people who probably arrived from Africa about a million years ago. The key find in this chronology was made in 1994, when a human premolar tooth, with a primitive, African appearance very like the teeth of the Nariokotome boy, was found by a team led by Arsuaga's colleagues – Eudad Carbonelli and Jose Maria Bermudez de Castro, both Madrid-based paleontologists. The tooth was located under a layer of sediment dated at 780,000 years old.[4] The find triggered intense excitement, and other fossil fragments were soon dug up, revealing part of a child's skull with a graceful, double-arched brow ridge, again like that of the Nariokotome boy. It was an important discovery because it pushed back the date of mankind's presence in Europe by several hundred thousand years. In addition, the Dolina fossil provided strong support for the idea that *erectus*-like settlers had wandered from their African homeland and colonized Europe. (Arsuaga has given a special, separate name to these late *erectus* settlers: *Homo antecessor*.)

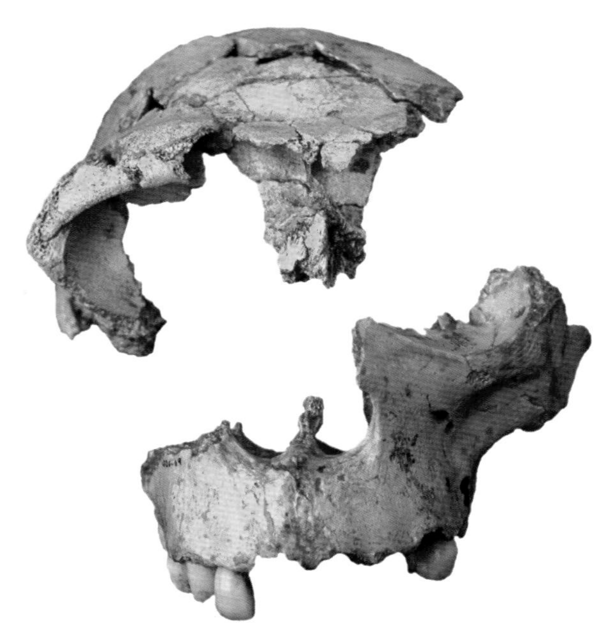

A skull of one of the first Europeans unearthed at Gran Dolina. Dating indicates that it may be more than 800,000 years old.

The next part of our evolutionary story is less clear, however, for Dolina has yet to produce clues about our subsequent development in Europe. So we must look elsewhere – in this case, to the green, leafy lanes and quiet villages of West Sussex in England, where scientists have also painstakingly unraveled ancient lives, although this site, Boxgrove, is very different from any at Atapuerca. There are no caves and few human fossils, only open quarry and a vast arsenal of stone weapons and tools.

Half a million years ago, an ancient sea lapped against limestone cliffs at Boxgrove. Rhinoceroses, horses, giant deer, and other animals ambled along the beach under the watchful eyes of human hunters. The weapons that were abandoned by these people – stone handaxes, flint carvers, and spearheads – have been carefully amassed by archaeologist Mark Roberts, who spent more than a decade living at Boxgrove until the dig was closed in 1996.

Crucially, the way these implements were used suggests that a critical change had taken place in mankind's activities, which now showed signs of real intelligence, says Roberts.[5] For a start, analysis of the remains found by the archaeologist – and by the hordes of willing volunteers who labored for 11 seasons at Boxgrove – shows no marks made by stone tools on any of the remains of the small mammals and birds that were found there. By contrast, bones from large animals – rhinos,

Some of the layers of soil laid down at Boxgrove in the millennia since *Homo heidelbergensis* first ruled the landscape there half a million years ago.

horses, hippos, and others – reveal a patina of man-made slashes. Significantly, these are found *under* the chew marks of other carnivores. From these discoveries, scientists draw two clear conclusions: first, humans were targeting only large animals; second, they were getting to their prey first, suggesting that they were not scavenging but were bringing down deer and horses in their own right as fully fledged hunters. A species that had made a living by stealing the leftovers of other carnivores' dinners in Africa had, 1 million years later, evolved into the master of Boxgrove's beaches and cliffs.

Indeed, cut marks on rhino bones reveal a complete sequence of sophisticated butchery that was obviously done in an unhurried manner – there was no fearful looking over shoulders, no anxious watching for the approach of rival predators. "They didn't wait until a rhino dropped dead of old age, or steal it from other hunters, such as lions, hyenas, or wolves," Roberts says. "Their prey was carefully selected. In one part of the dig, we found four butchered rhinoceroses. Each would have weighed 1,500 lb (675 kg), a magnet for other predators. Yet each carcass was skillfully cut up. Fillet steaks were sliced from the spine, and the bones were smashed to get out the marrow. Only hunters who were in total command of their patch could have done that."[6] These were not simple-minded scavengers, in other words. They were careful planners, men and women who were thinking in highly complex ways. Nor is it just the complexity of the hunt that reveals the sophistication of their way of life. What also impresses Roberts is the way these people seem to have divided and shared the meat among themselves. And that means only one thing, he says. "For this kind of hunting, involving stalking and ambushing, and for ensuring the proper distribution of meat, speech would have been essential, I believe. They had to have been able to exchange quite complicated ideas rapidly and efficiently."[7]

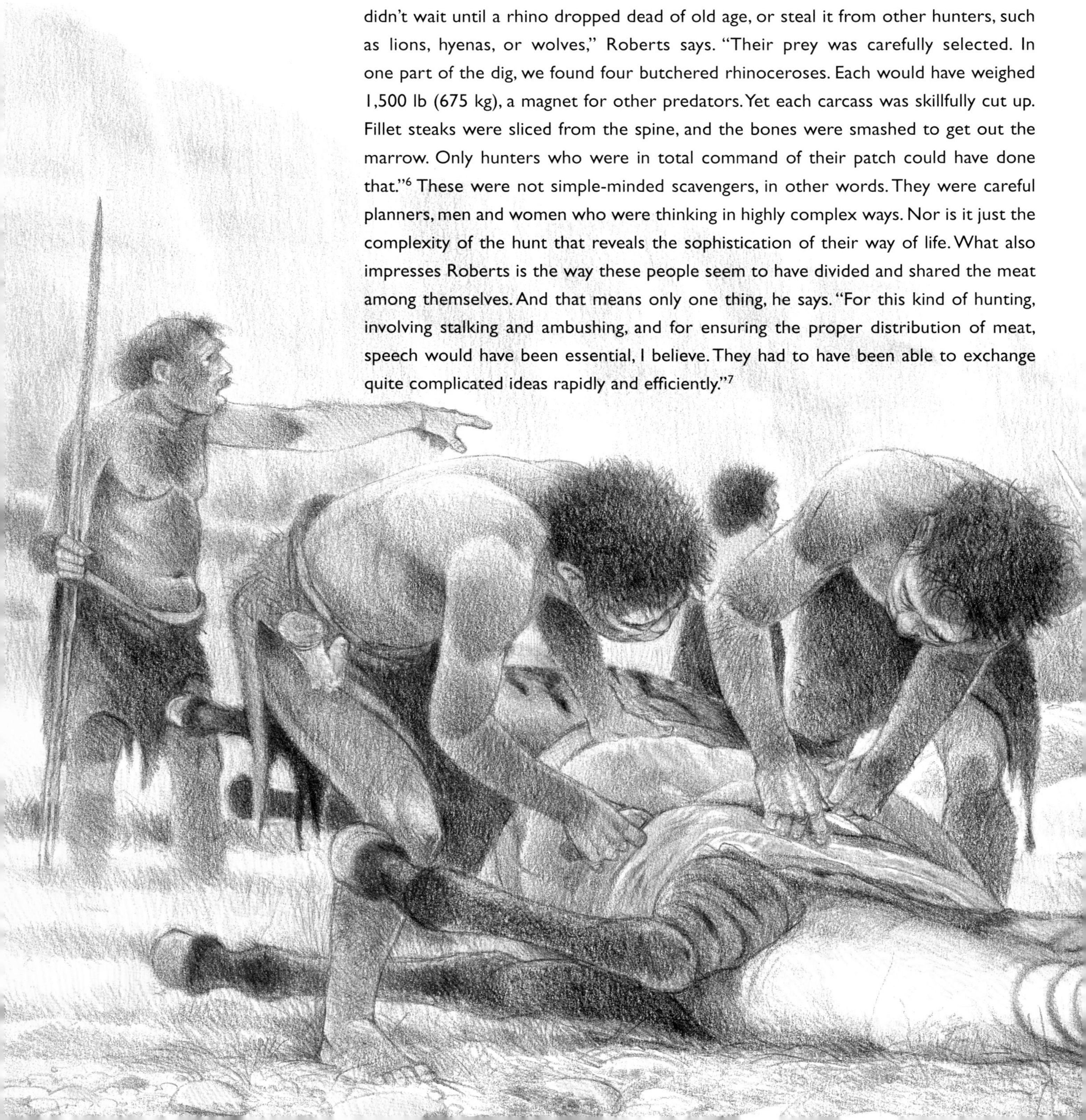

The idea is fascinating but, unfortunately, impossible to substantiate. The spoken word leaves no fossil or archaeological record, making the linguistic ability of our predecessors one of the most difficult and contentious issues of human evolution. Most anthropologists accept that our predecessors must have communicated in some relatively sophisticated way. How close the conversation of Boxgrove folk would have been to the language skills of *Homo sapiens* is a different matter, however. The issue is summed up by paleontologist Ian Tattersall, who believes that, with the arrival of hominids of this period, the element of human language potentially enters the picture. "For the first time, we encounter a skull base anatomy that suggests its possessor had all the peripheral equipment necessary for speech production." He sees little in the archaeological record to support the idea, but acknowledges that "however individuals of this species communicated, they did so in a quite sophisticated way."[8]

In the meantime, let us consider a more pressing question: who exactly were these ancient hunters? Scientists have recently given them their own name: *Homo*

Homo heidelbergensis dominated the Boxgrove region – despite the presence of lions and other competing carnivores – and tracked and ambushed their prey with cunning and careful stalking. Once heidelberg hunters had killed their quarry, they cut it up and distributed it (below) with the confidence of predators in complete command of their terrain. This artist's impression shows a scene from life at Boxgrove.

One of the most sophisticated tools discovered at Boxgrove, this hammer (top) would have been carved from the antlers of the now extinct giant deer *Megaloceros dawkinsi* (above), which is shown here in a reconstruction made by the Natural History Museum, London. It is believed the hammer was used for making stone tools.

heidelbergensis, after a large, chinless lower jaw – believed to be about 500,000 years old – that was found at the Mauer sand quarry near Heidelberg, Germany, in 1907. (*Heidelbergensis* is still occasionally known by its old name, archaic *sapiens*.) *Heidelbergensis* fossils have since been uncovered at several sites in Africa (at Bodo in Ethiopia and at Kabwe in Zambia, for example) and Europe, as we shall see. It was, we believe, a direct descendant of *Homo erectus*, although it was also, very clearly, a different class of human, as is revealed in perhaps the most striking of all Boxgrove's artifacts: a hammer, worn from overuse, that had been carefully carved from the antler of a giant deer called *Megaloceros dawkinsi*, which became extinct 500,000 years ago. Roberts reckons the hammer was used to chip and shape dozens of stone handaxes, before being discarded or lost. Microscopic examination showed that the ancient mallet is still embedded with tiny flint fragments, leftovers from a time when it was used to knap stone tools from a large chunk of Boxgrove flint.[9] Most probably, the hammer had been carried around, owned – most likely treasured – by one individual, giving Roberts's find a unique status: mankind's oldest-known possession.

For archaeologist Clive Gamble, who is based at Southampton University, the antler hammer has particular significance. "This tool was carried about by someone who was ready to use it at any time. Its owner was a prehistoric boy scout, a person who knew to be prepared. These people must have had clothes, it was so cold then, and if you have clothes, it is not a major feat of imagination to envisage them having some sort of pocket. The owner of the hammer would have kept it on his, or possibly her, person, ready for use whenever needed. And that implies a real improvement in thinking."[10] *Erectus* people probably moved materials around in preparation for specified acts; but their successors, the heidelbergs, looked as if they were always ready to make a new tool and to kill whenever the chance came along. They were equipping themselves for different eventualities.

Further evidence of this hunting prowess has been found at Schöningen, near Heidelberg. At the site of a vast opencast mine, scientists uncovered five superbly crafted spears that have been dated at about 400,000 years old. In those days, says Hartmut Thieme of the Institute for the Preservation of Historic Monuments, there would have been a lake at Schöningen. Animals that came to drink at the shoreline might have been ambushed by hunters armed with spears. "Try to imagine the scene 400,000 years ago," says Thieme. "A lovely lake with herds of horses drinking at the shoreline. Humans hiding in the bushes. At the right moment, they leaped out, throwing spears. Since the horses wouldn't run into the water, it was easy to catch them."[11] The carpentry involved in making these weapons was startlingly sophisticated, reckons Thieme. The ends of the spear shafts seem to have been expertly split open, making perfect joints in which to insert stone spearheads. In addition, the carpenters knew to make their spears from slow-growing trees, which would have provided the strongest and hardest wood. If Thieme's interpretation is correct, these people were practicing tool-making of unexpected sophistication.

Archaeologist Hartmut Thieme, who has found evidence at Schöningen in Germany that 400,000 years ago humans were making sophisticated spears from the branches of slow-growing trees, thus providing themselves with the strongest, hardest wood for weapons.

Until the 1990s, when the key discoveries were made at Boxgrove and Schöningen, scientists had no evidence that humans of this period were capable of stalking or planning hunts. The work of Thieme and Roberts has now transformed that uncertainty into accepted fact. In the latter's case, it was to be a fitting reward for 11 years of privation, living at the edge of a disused quarry in an old milking parlor, with a bed made from wooden pallets and neither electricity nor running water for most of the years that he lived there.

But in spite of the breakthroughs made by Thieme and Roberts, many aspects of the heidelbergs remain tantalizingly mysterious. They dominated Boxgrove and

The Boxgrove tibia, the shinbone of a mighty individual who lived 500,000 years ago. It has twice the bone tissue of a modern human's tibia, and the large ridges along its back suggest that its owner's leg muscles must have been extremely powerful.

Schöningen, and were among the best handaxe-makers the world has seen, says Gamble. But they left behind no indication that they built shelters or permanent hearths. "We found one small area of Boxgrove that had a layer containing crumbs of charcoal, and a flint – blotchy pink and purple – that looked as if it had been baked in a fire," says Roberts. "But whether it was a hearth or a natural fire of some kind, we couldn't tell. Frankly, the jury is still out about whether or not the heidelbergs were regularly, systematically using fire – though I believe they did. We didn't find one in the area we excavated, but then we only excavated the part of the countryside where they hunted and butchered – the bit below the cliff. They probably took their prey up to the top once they had cut the meat off. In other words, we only found the killing floor, not the kitchen."[12]

For answers to questions about whether these people were completely nomadic or had begun to build homes, we have to look elsewhere: in particular, to Bilzingsleben in Germany. Here, Dietrich Mania of the University of Jena dug up three mounds of bone and stone and, in the center of one, found a long elephant tusk. He interprets these structures as dwellings, with the tusk acting as a center post, and has dated them at between 412,000 and 320,000 years old.[13]

That is startling enough. But Mania goes further. He claims that another, wider mound that he unearthed at Bilzingsleben contained signs that humans were now capable of abstract thought. As proof, he points to a slice of elephant shinbone criss-crossed with a series of regular lines that he interprets as graphic symbols – evidence of abstract thought. Such intellectual prowess would be the earliest attributed to a human being. Mania's explanation is controversial, however, and many anthropologists believe symbolic thinking did not manifest itself until much later in our evolution.[14] As Gamble puts it, "Bilzingsleben is a very interesting site, but the majority view is that the huts are not proven, nor the evidence of symbolic logic."[15]

Another mystery is what the heidelbergs looked like. Scientists think they evolved from *Homo erectus* about half a million years ago, but did they retain the tall, African physique of *erectus* or had they adapted to the European cold by becoming shorter and rounder? Unfortunately, there are not enough fossils to give a clear picture of the heidelberg anatomy. For example, only a couple of molar teeth and a shinbone were found at Boxgrove. A *heidelbergensis* skeleton comparable to that of the Nariokotome boy would make all the difference, say paleontologists, although this has not stopped them from making some educated guesses based on existing material. For one thing, the Boxgrove shinbone is enormous, with at least twice the bone tissue of a modern human's. It also has large ridges along its back, which implies that the muscles attached to them were enormous and powerful. Its owner must have been about 6 ft (1.8 m) tall and very strongly built. In other words, *heidelbergensis* was probably as big as *erectus*, if not bigger, and shows little sign of having evolved the squat physique of later, cold-adapted European hominids. It seems that this species came out of Africa in a late wave of post-*erectus* settlers and was not the product of a lengthy evolution in Europe – although not every scientist agrees with this scenario.

However, most researchers agree on one thing: that the heidelbergs were intellectually superior to their predecessors. From fragments of skulls found at Steinheim in Germany, Petranola in Greece, and Aragu in France, we can estimate that their braincases had volumes of about 1100 cc – "within striking distance of the modern average," as Tattersall puts it.[16] Certainly, it is significantly greater than *erectus*'s 800–900 cc. "Boxgrove man would stand out a bit today but we would recognize him as human," says anthropologist Erik Trinkaus of the University of New Mexico.[17]

The increased brainpower may have given *heidelbergensis* the behavioral flexibility it needed to survive in an era of severe climatic upheaval. About 600,000 years ago, the world was passing through a series of savage cycles of cooling and heating that included six major ice ages, says Richard Potts. Our planet's higher latitudes would have been coated in glaciers at times, turning Boxgrove and Schöningen into grim, unfriendly landscapes. At other times, they would have been positively balmy. This vacillation between subtropical and arctic conditions would have stretched and pushed the heidelbergs' survival skills to the limit. The result was "the emergence of the most advanced means of human social and mental problem solving," says Potts.[18]

So it was around this time that the last of the three key human attributes – bipedalism, tool-making, and big brains – really began to manifest itself. "We became versatile apemen when we walked, and more adaptable hominids when we made tools that gave us new sources of food," says Potts. "But when our brains really started to grow, as we can see happening with *Homo heidelbergensis*, this allowed our ancestors to be even more flexible to life's vicissitudes, to solve quite complex problems, and to form sophisticated social bonds that would have allowed them to pool their resources in the face of growing adversity. And at this time, there would certainly have been plenty of adversity. The climate went through cycles of intensifying extremes, and we responded accordingly – by becoming even more flexible in our behavior."[19] In other

MASTERS OF STONE

Mankind almost certainly used tools very early in its evolution, although it is most likely that this behavior merely involved the opportunistic exploitation of conveniently shaped stones or twigs. The real question is: when did early men and women first fashion implements to suit their needs? In other words, when did they begin to shape nature to their own ends?

This is a very difficult question to answer – for it is quite possible that our predecessors have always honed natural materials in order to turn them into utensils, albeit primitive ones. Pioneering studies at Gombe in Tanzania by the distinguished primatologist Jane Goodall have shown that chimpanzees will bend, twist and shape branches and twigs to make implements that they use for various purposes, such as collecting termites (a favorite chimp food) and scooping them into the chimps' mouths. The common ancestor of man and ape might, therefore, have used tools in this way. However, since such implements would have been constructed from biodegradable materials, we have no record of them.

What is certain is that by 2.6 million years before present, early humans had begun to fashion primitive

devices from true, enduring materials, such as rock and lava. At sites in the Hadar region of Ethiopia and from around Lake Turkana in Kenya, scientists have uncovered quartz pebbles that had been shattered to create sharp-edged implements. Scrapers, pounders, and choppers were made from large pieces of stone, and were probably employed to break bones and process marrow, while the flakes that flew off during knapping would have been used to cut hide or meat.

This basic technology has evolved, very slowly, over the millennia. For example, stone hammers and anvils would have helped our ancestors to crack open nuts, while animal bones would have been exploited in digging up tubers and roots. Mankind's growing prowess in tool manufacture eventually reached a pinnacle when men and women were able to create finely made bone and stone implements that could be used for a variety of different specialized applications. Indeed, in some parts of the world, such as New Guinea, tribesmen still manufacture such implements from basic cores of rock. In a sense, stone tool-making is mankind's most enduring technology.

FAR LEFT: A recreation of *Homo erectus* making stone tools. Suitably sized pebbles would be carefully collected and then formed into implements for various purposes. LEFT: A chimpanzee scoops termites from a nest using a branch that it has specifically shaped for this purpose. ABOVE: A core from which stone tools are chipped.

words, guile and brainpower played an essential role in the survival of *heidelbergensis*. Social interactions would also have become important, with increasing reliance being placed on understanding the intentions of fellow members of the group. Communication of a fairly sophisticated nature would certainly have gone on. "With the increased verbal communication, you would also get the downside of language, of course – the ability to cheat," says Leslie Aiello. "Following that, you would have had to evolve ways to detect such cheats, and so the ever-increasing complexity of human behavior would have snowballed."[20]

The heidelbergs would also have relied on their mighty physiques, as Roberts and colleague Michael Pitts argue in their book *Fairweather Eden*. "The heidelbergs and their descendants in Europe put everything into the hunt. Combining their stone-working skills with sheer brute force, they could outsmart both the toughest animals and the harshest climate." We shall examine the consequences of this adaptation later on. For the moment, let us confine ourselves to the question of how *heidelbergensis* compares – intellectually, socially, and spiritually – with *Homo erectus* and, in particular, with the vision of the latter hominid that was outlined by paleontologist Alan Walker in Chapter Three. As you may recall, Walker believes that *erectus* lacked the spark of human consciousness, although many other paleontologists disagree. They see no reason to so deride *erectus*, although it is difficult to draw conclusions from present evidence. But do we have better evidence for *Homo heidelbergensis*? Would we have seen anything more than just a dull glow in its gaze? Roberts thinks so. "It would be like looking into the eyes of a drunk Glaswegian at midnight: definitely something going on, but you can't be sure what it is."[21] It sounds impressive in a perverse way. But was it enough to ensure the survival of the lineage?

The answer to that question is intriguing because it has great bearing on our understanding of the appearance of our own species, *Homo sapiens*, and it takes us back to where we started this chapter: Atapuerca. As we have seen, there is good evidence that *erectus*-like humans settled in this region about 800,000 years ago. Then, at Boxgrove and Schöningen, we find remains of their successors, *heidelbergensis*, dated at 500,000–400,000 years old. But what happened after that? Paleontologists could only guess – until a chance find sent them stumbling into the underground labyrinth of Atapuerca, and La Sima de los Huesos.

The cavern system was once a hibernation den for cave bears, whose teeth and bones littered its floors. In the 19th century, local men would show their daring by

OPPOSITE: Inside the Pit of Bones, Spanish paleontologists excavate 300,000-year-old fossils in a tiny cave at the end of a maze of potholes and caverns.

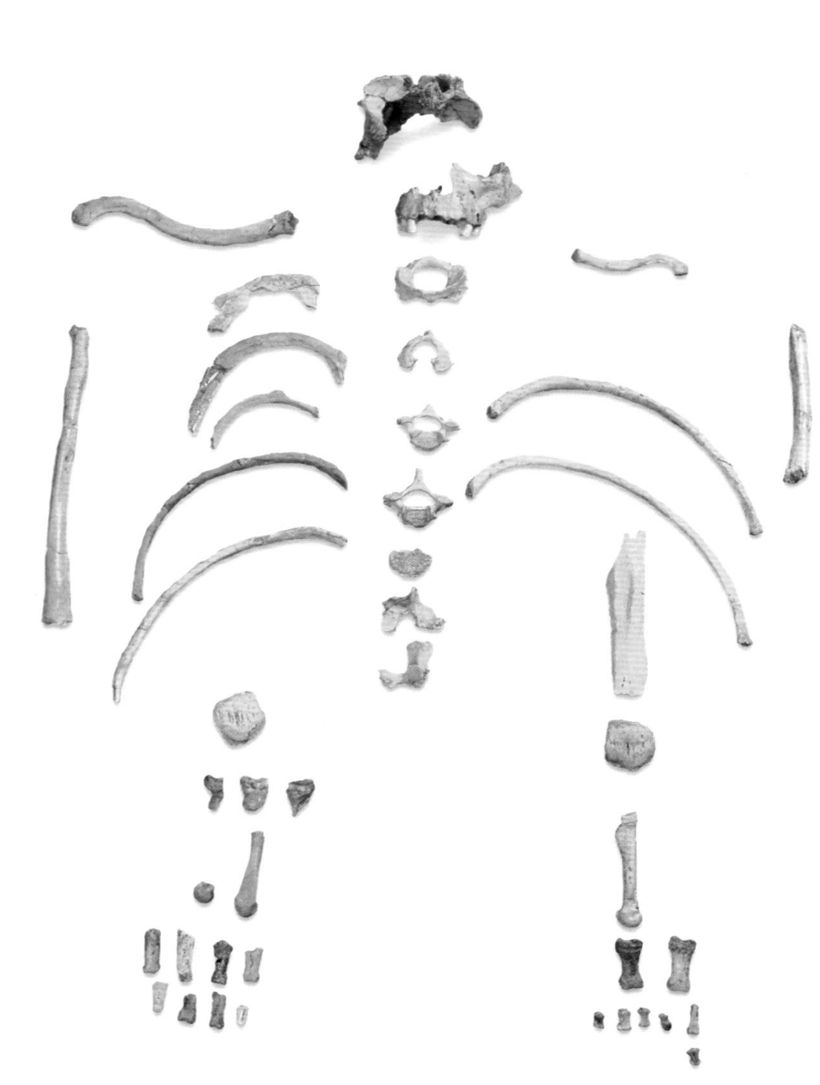

An 800,000-year-old partial skeleton found at the Gran Dolina site in northern Spain. Its great antiquity came as a surprise to some scientists, who had previously believed that the first settlers arrived in Europe around 500,000 years ago.

crawling into the tunnels to retrieve this ursine detritus to make necklaces for their sweethearts. Then potholers began to map the caverns systematically, and uncovered the Pit of the Bones. At the bottom they found mounds of cave-bear skeletons, in which a Madrid researcher, who was studying these extinct creatures, also found fragments of human teeth and skulls.

Juan Luis Arsuaga, then a postgraduate paleontology researcher, thought the remains looked promising. "Unfortunately, the pit was filled with rubbish left behind by potholers, and the sediments and layers were all jumbled up," he recalls.[22] So Arsuaga recruited fellow members of his geology faculty's rugby team to help clean up the pit. Every summer, they would slither into La Sima, fill their backpacks with debris, and lug them to the surface. "I sold it as a challenge," he says. "I was sure we would find something really good. We kept finding tiny finger bones, little fragments that I called *galletas* – biscuits. Normally, these are the first pieces to be washed from an excavation. Their presence meant there had to be real treasures there, possibly whole skeletons."[23] In 1990, excavations at the pit began in earnest and quickly revealed a fairly respectable haul of human skull fragments and bones, all dated at about 300,000 years old. Two years later, one of Arsuaga's team found a fragment of bone that refused to be dislodged. After carefully scraping, a large piece of skull was unearthed. By the end of that year's dig, the team had pieced together two complete ancient human braincases. Then, on the last day of the season, Arsuaga and his assistant, Ignacio "Nacho" Martinez, made a final visit to tidy up the pit. "Nacho kept nagging me to do a final

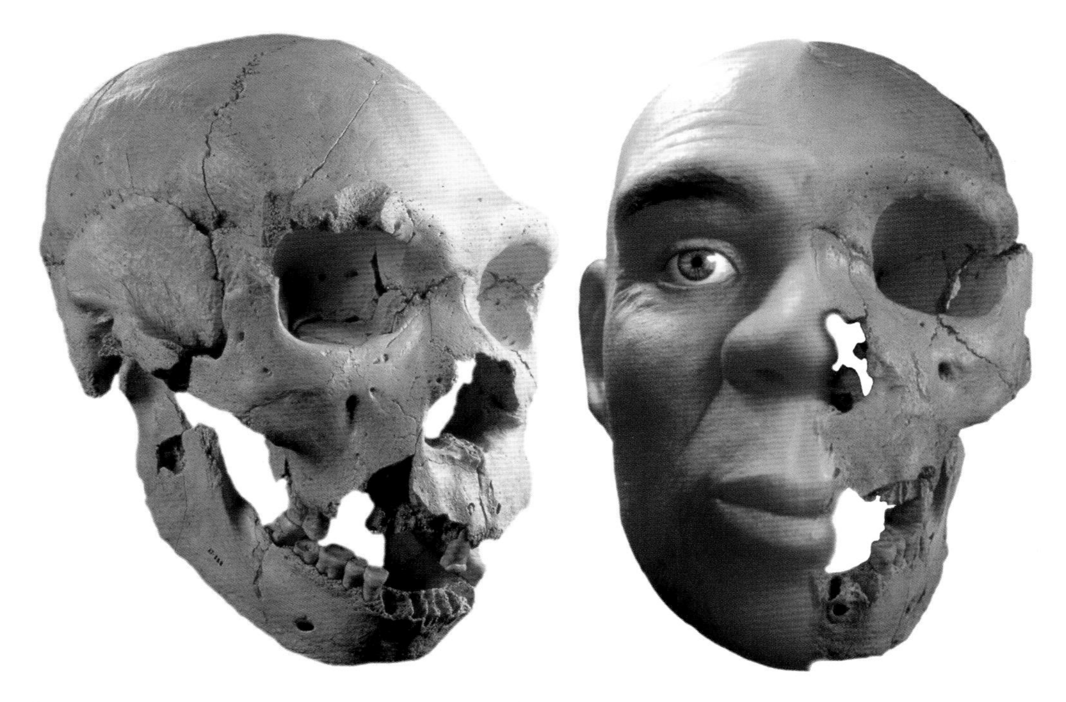

A complete skull (above left) discovered by Spanish paleontologists in the Pit of Bones. From this skull, a reconstruction was made (above right), revealing an individual who still has the heavy browridge of *erectus* but whose braincase and skull are of a different shape.

bit of excavating, and I kept saying no, what happens if we find something? We could be here for months. In the end he persuaded me, and we started scraping away at some sediment. That is when we found the face. There was a whole front of a skull in the mud. We took it to the surface – and it fitted exactly with one of the braincases. We had got our first complete skull."[24]

Since then, researchers have been overwhelmed with fossil goodies from the pit, although they are still left with a major headache: how did dozens of ancient bodies end up stuffed down a Stygian pit inside a pothole labyrinth? It is a significant question, for the answer may reveal how far humans had evolved toward acquiring true intelligence. No stone tools or other artifacts have yet been found in La Sima, so it seems unlikely that the bones came from a cave dwelling that collapsed on its occupants. The idea that the bodies may have been dragged into the cave by predators such as lions – as some scientists have proposed – is also dismissed because there are no remains of such species mixed with the human bones. (The bear bones were deposited much later.)

Instead, Arsuaga believes bodies were deliberately thrown down La Sima. "When a person died, their relatives carefully carried them to the pit and flung them in. Of course, there was probably another, more convenient route to the pit, a tunnel

MAKING HEADWAY

The brain of *Homo sapiens* is one of the more remarkable features of our species, along with our upright gait and complex spoken language. It is greedy in its demand for resources, however, and consumes 18 percent of a person's energy budget even though it represents only 2 percent of his or her body weight. Another illustration of how our brains dominate our anatomies is to imagine an ape the same size as a man or a woman. Its brain would still only be a third of the size of a human being's. We are, in short, a bunch of brainy apes.

What is unclear is whether the threefold enlargement in our brain size since the beginning of our evolution represents a steady accretion of capacity or was a matter of sudden increases associated with special phases in our evolution. In other words, was our brain growth a smooth business or did it involve particular episodes in our past – the first tool-making, the first meat-eating, and the rise of complex societies? Did these events trigger rapid jumps in cranial volume, only to be followed by longer, relatively slow periods of expansion? Based on evidence from the fossil record, most scientists believe that this scenario is the right one.

Australopithecines
All members of the *Australopithecus* clan – *anamensis*, *afarensis*, and *africanus* – had brains that were little bigger than those of chimpanzees or gorillas. On average, they had a capacity of little more than 400 cc. So, while they had evolved anatomically to walk on two legs, intellectually they were still pretty much on a par with their ape predecessors.

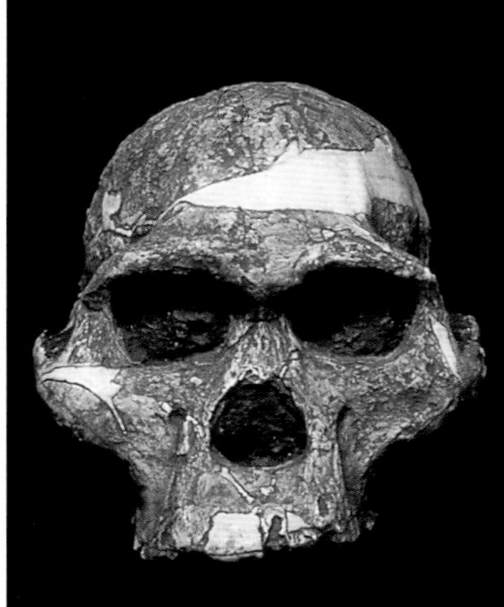

Homo habilis
Habilis is the first of our lineage to show signs of significantly increased cranial capacity – about 650 cc. In addition, the many stone tools found near the remains of this species suggest that this was the first true member of humanity, a moderately big-brained tool-maker. Recently, some scientists have challenged this idea and have suggested that *habilis* should be "downgraded" to australopithecine status.

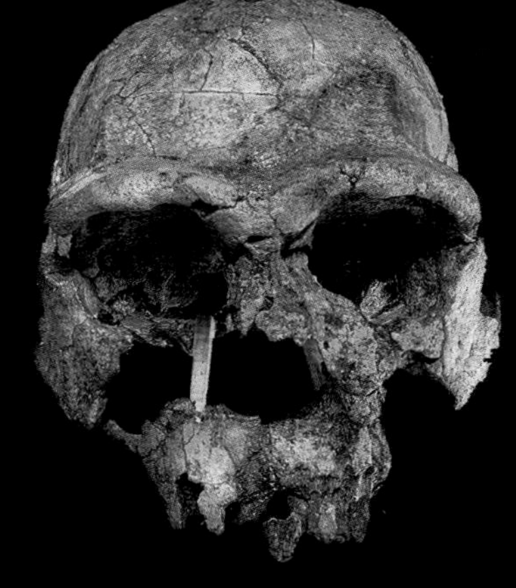

Homo erectus

Erectus represents another clear stage in the intellectual evolution of modern human beings, with a brain that had a capacity of about 850 to 900 cc. The species' intellectual prowess is the subject of controversy, however. Paleontologist Alan Walker argues that its intelligence would have been unimpressive. Others point out that with a brain twice the size of a chimp's, *erectus* could certainly not be described as an evolutionary dunce.

Homo neanderthalensis

The Neanderthal brain, at least in terms of simple capacity, represents the culmination of the cranial evolution of our species. At over 1300 cc, its brain is as big as they come for hominids, and in some cases is actually larger than that of *Homo sapiens* – although this effect may have been a result of Neanderthals having larger bodies, which tend to need increased neurologic control mechanisms.

Homo sapiens

The brain of modern man is distinguished by its large forehead, a sign that the frontal lobes of the brain had taken the brunt of the increase in our cranial capacity. Our brains are also narrower and higher than those of our predecessors, and we have an average brain size of between 1200 and 1600cc, three to four times that of our earliest ancestors.

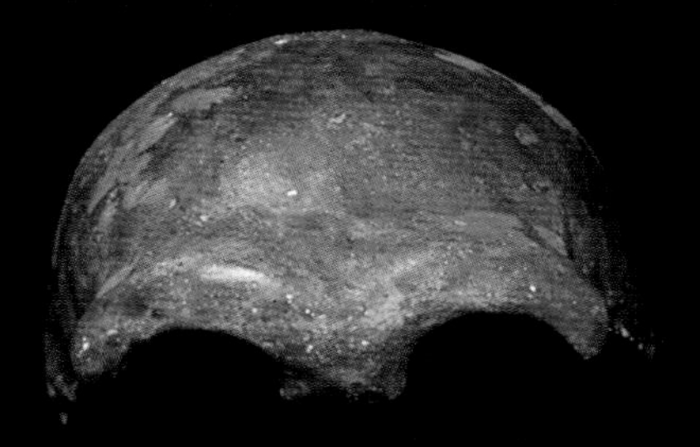

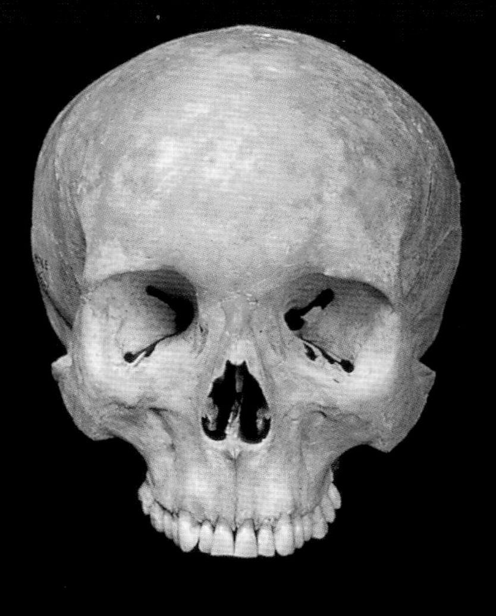

that has since collapsed."[25] Such disposal suggests that the Atapuercans in some way glorified their dead or possibly even believed in an afterlife. Three hundred thousand years ago, this was a special place, designated only for humans. "Spiritual life started when people first realized that everyone dies. Animals have no concept of death, unlike *Homo sapiens*. When our ancestors realized that death was inevitable, something changed inside them. That was the terrible consequence of increased brainpower. I believe that ancient Atapuercans reached that state, and that is why their bodies ended up in the Pit of Bones."[26]

It is an intriguing interpretation, although some scientists remain sceptical. "If Atapuercans were throwing their dead relatives into the pit, we should see complete skeletons being pulled out of La Sima," points out paleontologist Peter Andrews of the Natural History Museum, London. "But we do not. Teeth belonging to 32 different individuals have been found there, and if bodies were being thrown whole into the pit, we should then expect to find 64 femur bones from their legs. In fact, we only find 20. And the same goes for the rest of their anatomies. There are far too many missing pieces."[27]

Instead, Andrews argues that some other agency must have been involved in amassing La Sima's spectacular skeletal assemblage. He points to chew marks left by foxes, and the odd indentation left by the teeth of a large carnivore, possibly a lion. These animals may have been the creators of La Sima's horde of fossils – although Andrews acknowledges that this still begs the question of why we find only human bones down there. Dr Yolanda Fernandez-Jalvo of the Museo Nacional de Ciencias Naturales, Madrid, argues that the bodies may simply have been left at one of the cave's long-lost entrances. Foxes or lions may then have dragged away pieces of corpse, and some slithered down La Sima. This explanation provides a more prosaic reason for the pit's contents, but it still suggests that humans were selecting their dead for special treatment and were, therefore, capable of spiritual awareness. "I don't think it is far-fetched to attribute primitive religious beliefs to the Atapuercans," admits Andrews. "They were large-brained and most likely went on to evolve into species who certainly did practice some sort of religious rituals. However, I just do not think we have enough evidence from La Sima to be totally convinced as yet."[28]

BELOW: A pile of skulls and bones, as found in situ in the Pit of Bones.

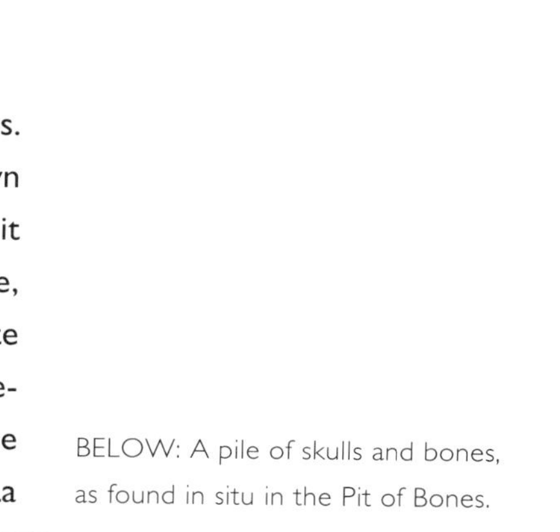

Intriguingly, nearly all those interred in La Sima were adolescents. Apart from a couple of adults and a child, the bodies were all those of teenagers, and they seem to have come from a closely related group who ended up in La Sima over a short period of time – "perhaps a year, no more than two or three," says Arsuaga.[29] In other words, these were young men and women who came from a single population and who would have known each other. Such genetic and temporal proximity speaks of some calamity, perhaps an epidemic – an idea supported by the observation that all the young folk of La Sima seem to have suffered some form of illness. One may have died from an infection that spread from a broken tooth; another must have been deaf, to judge from the growths that blocked his ear canals; and nearly all had tiny holes in their eye sockets – symptoms of a condition known as cribra orbitalia, which is linked to childhood malnourishment.[30] "There are signs of jaw and bone disease in most of the people of La Sima," says Professor Chris Stringer of the Natural History Museum in London. "It is possible that some kind of illness or epidemic killed them and they were dumped in the pit. However, until we find more evidence, we cannot really be sure what happened."[31] Certainly, life seems to have been very tough for these members of the *Homo* lineage. "If you look at their injuries, it is clear these folk must have suffered a high death rate, often dying in

OPPOSITE: An artist's impression of Atapuercans burying one of their dead in the Pit of Bones. The large number of hominid fossils found there has led scientists to believe that the Atapuercan people revered their dead and that they had, therefore, developed a spiritual life.

DIGGING UP THE PAST

The business of uncovering our prehistory has come a long way since archaeologists and paleontologists simply dug down until they had unearthed a single interesting object. Today a site is first surveyed in detail, the age and distribution of its soil and rock sediments carefully noted, before it is divided into sections using string to mark boundaries of even sizes. Only then can the delicate business of excavation begin – with researchers and volunteers using brushes or scalpels (not trowels or spades, which could damage delicate fossil remains) to scrape away soil with deliberate, painstaking care. It is an extraordinarily laborious business, made worse by the fact that it is often undertaken in searing heat – because most excavations are carried out during the summer when academics are on vacation.

All finds, no matter how small, are then marked and recorded on charts. Even the dirt that is scraped away is saved and sieved to ensure that not a single fragment is overlooked – it is another arduous, tedious, and unpopular job, but an essential one. In this way, every single object, however small, and every scrap of data is exploited to build up a picture of just what kind of lives our ancient predecessors lived. From such fragments scientists can tell what kind of creatures were hunted and killed at a human settlement, what kind of hearths were made, how people distributed food, and other details of their social organization.

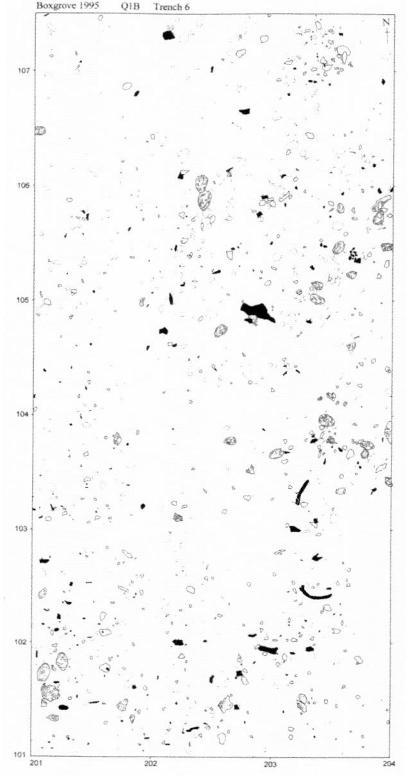

ABOVE: A record of finds from Boxgrove showing rhinoceros bones (in black), handaxes (the shaded, teardrop-shaped objects), and flints. It is essential to record not only *what* is found, but also where, since much can be deduced about our ancestors' behavior from remains that are found together – butchered bones with tools indicate that heidelbergs may have brought their kill to this spot to be carved up.
LEFT: An excavation site in East Turkana. Note the string dividers that are used to mark off areas of the site to ensure meticulous recording of each find and its precise location.

childhood or early adulthood," says Leslie Aiello.[32] "Many groups would have died out, while others would have had to struggle hard just to replace their numbers, never mind expand." This twig on the bush of life was in constant danger of being pruned, in other words.

As for the anatomy of the individuals found in La Sima, analysis shows that they were tall, with big noses, no chins, heavy brow ridges over their eyes, braincases of varying sizes, and sturdy physiques. They retain some *erectus* features, but they also show distinct signs of having adapted in some ways to their European homeland and its colder climate. The variation in brain size may seem puzzling, given that the people of La Sima were probably closely related. But, as Stringer points out, "If you look at groups of people today, you see lots of variation in head size: some have big heads, some small. It was probably the same then."[33]

So the people of La Sima were tall, like the heidelbergs, and had not yet begun to evolve shorter, rounder physiques in response to the cold. On the other hand, they did display other critical anatomical changes. For example, the middle of their faces projected quite strikingly and would have made their large noses even more prominent. This was probably an adaptation to the cold climate. With the entire middle face pulled out, inhaled air could be warmed before reaching the vicinity of the brain, which requires a stable temperature and blood supply. In addition, the front teeth found in La Sima all show extreme wear. Paleontologists believe this dental dilapidation comes about when a person uses the teeth as a third hand. In other words, they hold objects in the mouth while using the hands to cut or manipulate the object – effective, but hard on the teeth. Most probably, they ate like Inuits, clenching pieces of meat between their jaws and then hacking at them with knives. And, just like Inuits, they did so in a way that left a pattern on their teeth that showed they were predominantly right-handed. It is also clear from dental studies that the heidelbergs at La Sima used toothpicks to clear their mouths of unwanted meat.

These features are all highly characteristic of a lineage that would appear in Europe around 250,000 years ago, not long after the people of La Sima met their fate: the Neanderthals – perhaps the most controversial of all our predecessors. We shall see why the Neanderthals provoked such contention in the next chapter. What is important at this stage is to appreciate what La Sima is telling us. *Heidelbergensis*, which probably evolved from *erectus* approximately 700,000–600,000 years ago, was, by about 300,000 years ago, shedding many (but not all) of the trappings of its African origins. It seems that the people of La Sima were being transformed into a new type of hominid – the Neanderthal – which was to be the dominant form of human life on the continent for the next quarter of a million years.

At least, that is what was happening in Europe. Elsewhere – probably somewhere in central Africa – *heidelbergensis* would turn into a very different species from *Homo neanderthalensis*, although one that was just as remarkable. In other words, *heidelbergensis* probably stands where two branches of man's evolution divide. In Europe, one branch would evolve into Neanderthals, and at Atapuerca we have a snapshot of that transformation in progress. In Africa, the other branch would produce a subtly different species of human being. Two hundred thousand years later, the two – Neanderthals from Europe and *heidelbergensis*'s African successors – would meet. The outcome would be dramatic and of profound importance to us all.

CHAPTER SIX

FEARSOME CREATURES

"Nothing of his face was visible but a mouth bordered by raw flesh and a pair of murderous eyes. His squat stature exaggerated the length of his arms and the enormous width of his shoulders. His whole being expressed a brutal strength, tireless and without pity. Hairy or grisly, with a big face like a mask, great brow ridges and no forehead, clutching an enormous flint, and running like a baboon with his head forward and not, like a man, with his head up, he must have been a fearsome creature." Fearsome is putting it rather mildly. Utter terror, not alarm, would be a more appropriate response on meeting the owner of such attributes. These words would do credit to a genocidal alien, although the writer – H.G. Wells – was actually describing a Neanderthal, the species whose relatively recent transformation, from African interloper into fully fledged European, was captured so dramatically in the bones of Atapuerca, as we have just seen.

Wells's prose – from his 1921 short story *The Grisly Folk*[1] – makes the Neanderthals sound like atavistic killers and suggests that the author, normally the most humane of writers, had a certain distaste for them. In fact, Wells was just one of a multiplicity of literary figures and scientific authorities who have taken a disapproving view of the species. But Neanderthals have always had a bad press. Ever since they were discovered in 1856 – by workmen quarrying lime in a cave in the Neander Valley near Dusseldorf, Germany – they have suffered general vilification and considerable misunderstanding. These were half-men – deformed creatures – and could surely have

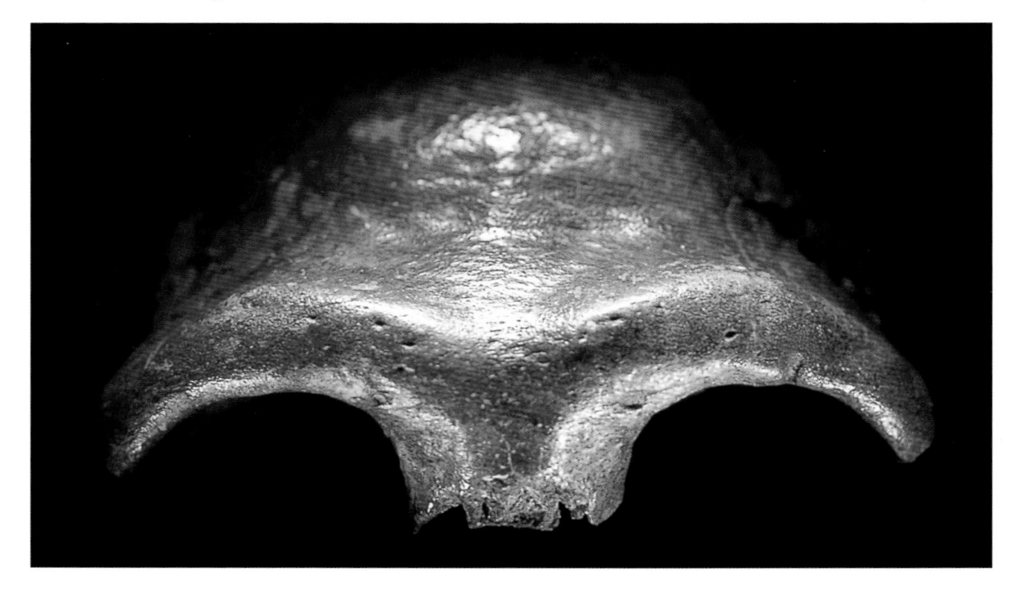

The skullcap of the man from the Neander Valley, Germany. Discovered in 1856, his bones were first thought to be those of a northern "savage," or a diseased horseman, or even a dead Cossack. Scientists now know that it belonged to a member of *Homo neanderthalensis*.

had no bearing on the story of modern humanity, it was argued. (There is even disagreement about their name. *Tal* means valley in German, but was spelled *thal* in the 19th century. As a result, some scientists prefer the word Neanderthal, while others use Neandertal. We shall stick to the former spelling in this book.)

It must be said that their remains do look odd. The quarrymen who discovered them were digging out mud from a newly blasted part of the quarry when they uncovered some ribs, part of a pelvis, and some arm and shoulder bones of a large, humanlike animal. They also found a skull that possessed a low, glowering brow ridge, and thighbones that were thick and curved. One of the quarrymen thought it was a bear's skeleton and told local schoolteacher Johann Karl Fuhlrott. A natural historian, Fuhlrott examined the bones and realized that there was something special about them, so he passed them on to the anatomist Professor Schaafhauson, who presented the remains at a meeting of the Lower Rhine Medical and Natural History Society in Bonn on February 4, 1857. Schaafhauson believed the bones were the remains of an ancient race of northern European savages – one of the barbarous tribes whose "aspect and flashing of their eyes" had terrified even the Roman armies.[2]

Rudolf Virchow, a German pathologist, examined the bones and announced that the peculiarities of the Neanderthal skeleton had been caused by rickets and nothing else. Laughable, said German anatomist F. Mayer. Those bent leg bones were not the result of disease; they marked the owner as a horseman whose damaged elbow showed he had been injured in battle. He must have been a Cossack cavalryman who had penetrated Prussia in 1814 in pursuit of Napoleon's retreating army, had suffered a sword injury and had crawled into the cave to die, said Mayer. It was a dramatic idea, if nothing else, but it did not convince the great British biologist Thomas Huxley, who took delight in pointing out the more basic flaws in Mayer's reasoning. How had the dying man managed to climb a 70-ft (21-m) precipice and bury himself after death, and why had he removed all his clothes and equipment before performing these strange stunts, Huxley asked.

In the absence of intelligent answers, Huxley concluded that the owner of those strange bones had not been quite like other men and women. The Neanderthal had odd, apelike characteristics, said Huxley, but he was definitely a member of the genus *Homo*. His brain size, which was well within the range of our own, made this clear. In the end, William King, an Irish anatomist, proposed that this was an ancient human, related to, but biologically different from us. On that basis, he named the first

BELOW: One of the more unsympathetic depictions of Neanderthals, portraying the character as a hairy, brutal, apelike thug.

distinct ancient human species to be discovered, *Homo neanderthalensis*.

The latter years of the 19th century and the early years of the 20th brought more discoveries of Neanderthal remains, particularly in Belgium and France. One of these – a complete skeleton found by Catholic priests in a cave at La Chapelle-aux-Saints, southern France – was to play a decisive role in the story of this species. The bones of the "Old Man of La Chapelle" were passed to Marcelin Boule, an eminent French paleontologist, who described the skeleton in great detail in an enormously influential monograph that was published in 1911.[3] Boule placed the Old Man in a kind of halfway-house between apes and humans, giving him toes that could grasp, a bent-kneed shuffle of a walk, and a curved spine that would have produced a permanently stooped posture.[4] "The likely absence of any trace of a preoccupation with an esthetic or moral order accords well with the brutal aspect of the heavy, vigorous body," Boule wrote. This was a species in which one could see clearly "the predominance of the purely vegetative or bestial functions over the cerebral ones."[5]

Such scientific demonizing explains where Wells got his anti-Neanderthal prejudice, if nothing else – although Boule's damning analysis was based on some rather unfortunate errors, as we shall see. Part of the problem lay in the fact that the Neanderthals that were unearthed at this time were the most formidable-looking ones. The species, we now know, had first appeared in Europe about 250,000 years ago and had adapted to the continent's fiercely cold conditions, evolving large noses to warm their breath, rounded bodies to retain heat, and powerful physiques. Males were on average 5 ft 6 in (1.68 m) tall and weighed about 140 lb (64 kg), while females were about 5 ft 3 in (1.6 m) and weighed about 110 lb (50 kg). Compare this with the tall, cylindrical frame of the Nariokotome boy, who would have measured at least 6 ft 1 in (1.85 m) had he lived to adulthood, and you can gauge the impact of evolution in a cold climate: Neanderthals had evolved short, stocky bodies to minimize the loss of heat through the skin. However, the basic Neanderthal form was not fixed tightly to this average – it varied quite widely according to locality. The sturdiest and stockiest populations lived in northwest Europe. These were the fiercest-looking, most alien individuals, and they were the first to be studied. That, in part, explains scientists' original low opinion of the species. Researchers were looking at the

The complete skeleton of the Old Man of La Chapelle. His bent spine was initially believed by scientists to indicate an apelike stooped posture, but was, in fact, the result of severe arthritis.

NEANDERTHALS: FLOWER PEOPLE OR MUGGERS?

Neanderthals have always had a bad image problem. As anthropologists Chris Stringer and Clive Gamble state in their book *In Search of the Neanderthals*: "No other group of prehistoric people carries such a weight of scientific and popular preconceptions, or has its name so associated with deep antiquity and the lingering taints of savagery, stupidity, and animal strength."

Much of this vilification can be traced to a popular urge among scientists of the Victorian and Edwardian eras to downgrade other versions of humanity to place *Homo sapiens*, and in particular Western men and women, at the top of the "ladder" of evolution. It therefore suited their purposes to stress the "allegedly" innate bestial nature of the Neanderthal.

For example, an illustration in the *Illustrated London News* (opposite, left), which was based on Marcellin Boule's study of the Chapelle-aux-Saints skeleton, shows Neanderthal man as a kind of Stone Age mugger waiting around a corner ready to pounce on innocent victims and attack them with a club. Most other images of the era are similarly unflattering. Men and women of the species are invariably depicted in slouched, bent-kneed poses, usually carrying a weapon of some kind.

One exception was provided by the anatomist Sir Arthur Keith (who had scorned Raymond Dart for suggesting that the Taung child represented a human-like being). He commissioned artwork, also based on the Chapelle-aux-Saints skeleton, and this time Neanderthals were depicted in a fairly warm light. The picture shows a calm figure sitting quietly beside a fire (below, right). This was not a brutal killer. Keith's vision of Neanderthal man was an exception for the time, however, and it was not until the 1960s that science came to terms with the idea that Neanderthals were not grunting, club-waving caveman.

These drawings, all from the *Illustrated London News*, reveal the public's fascination with Neanderthals, and demonstrates that even in the early 20th century, views about the nature of the species were divided. Most depictions of the species were unfavorable, such as the picture (below, left) of the Chapelle-aux-Saints man, which paints him in a vicious, ugly light. By contrast, Sir Arthur Keith presented the species in a far fairer light, including a vision of the Chapelle-aux-Saints man as a homely, contemplative figure (below, right). Sir Arthur Keith was also responsible for commissioning the illustration (opposite) of a group of Neanderthals deliberately stampeding and attacking elephants and hippos.

most extreme specimens. By contrast, more lightly built Neanderthals had lived in warmer eastern Europe and western Asia. A collection of their remains was assembled by paleontologist Ales Hrdlicka, who was born in Bohemia but emigrated to the United States, where he became one of the founders of the science of paleontology. In the 1920s, Hrdlicka became a proselytizer for the Neanderthals, but was widely ignored.

It was not until after World War II – almost 100 years after the discovery of the first Neanderthal – that the species began to get anything like fair treatment from the scientific establishment. One of the most important acts in this reappraisal was performed by the US anatomists William Straus and A.J.E. Cave, who decided to reexamine the Old Man of La Chapelle, the Neanderthal that Boule had used to paint such an unflattering picture of the species. They concluded that the eminent French paleontologist's analysis was unsound from beginning to end – or, to be more precise, from head to toe. The Old Man's bent spine was not evidence of a primitive, stooped posture, but the result of severe arthritis. Even worse, it became clear that Boule had noted this fact but ignored it in his desire to malign Neanderthal man. Straus and Cave concluded that a Neanderthal was not much different from a modern human.[6] Indeed, from analyses such as those by Straus and Cave, it became clear that the Neanderthals' brains were actually, on average, slightly larger than those of humans today, measuring about 1200–1750 cc in adults (probably because they had larger bodies, which needed more neurologic controls). However, compared to *Homo sapiens*, their skulls were flatter on top, smaller at the front, and bulged more at the sides and back. There are clear indications that they were right-handed – their brains were slightly larger at the right frontal and left rear areas, as are those of modern humans. And, like their heidelberg ancestors, their teeth showed signs of being used to hold objects, probably while they slashed at them with stone knives. Most remarkable of all, however, was the Neanderthal nose, which had both prominence and breadth, its projection being accentuated by cheekbones that flared back on each side of the face. As with the heidelbergs, the large nose was probably an adaptation that helped warm inhaled air before it passed near the brain.[7]

These, then, were the first hominids to adapt fully to conditions in Europe. And here they flourished for more than 200,000 years, leaving evidence of their presence in sites ranging from Wales and Gibraltar in the west to Moscow in the north, Uzbekistan in the east, and the Levant in the southeast.

Amud, a vaulted limestone cavern in upper Galilee where Yoel Rak found the remains of a buried Neanderthal child.

Their remains reveal a sophisticated species, capable of religion, art, and social concern, although their most strikingly human acts involved their burial of their dead – a custom that was formerly considered to be the sole spiritual prerogative of *Homo sapiens*. Several Neanderthal graves have been discovered in recent times, including one found by paleontologist Yoel Rak at Amud, a limestone cave high above a riverbed in upper Galilee, Israel. On a dig in 1992, a student was scraping at a patch of ground near the cave's north wall when the dusty outline of a distinctive piece of bone appeared. "We started brushing away the dirt, slowly, slowly," recalls Rak.[8] "And there it was: a skull. It started with a skull. A very tiny skull, the size of my fist, started emerging. It was a very exciting thing." Gradually, the infant's skeleton was exposed within its resting place, where it had lain for 60,000 years. Importantly, the skeleton was relatively intact and appeared to have been laid in position with its arms pressed by its sides. In addition, the jawbone of a red deer had been placed over the infant's pelvis. "It was an intentional offering, although it isn't clear whether it was meant as food for the afterlife, or as some more symbolic gesture. However, this was a primary grave, and the baby was deliberately put in it," Rak says.

Over the years, paleontologists have amassed a better collection of Neanderthal remains than of any other hominid (apart from ourselves), and many of these remains have been found in excellent condition. It is now clear that the reason for this abundance of well-preserved bones is simply that these people were intentionally burying their dead. At Le Moustier in southern France, the body of a young Neanderthal man was found flexed as if asleep and sprinkled with red ochre. At nearby La Ferrassie, an entire family – a man, woman and four small children – were found buried together. And the Old Man of La Chapelle, it now transpires, had been lowered into a trench that had presumably been dug by his family or friends – the same individuals who must have cared for him after arthritis made him unable to obtain his own food.[9] So much for Boule's assertion that these people were little more than animals.

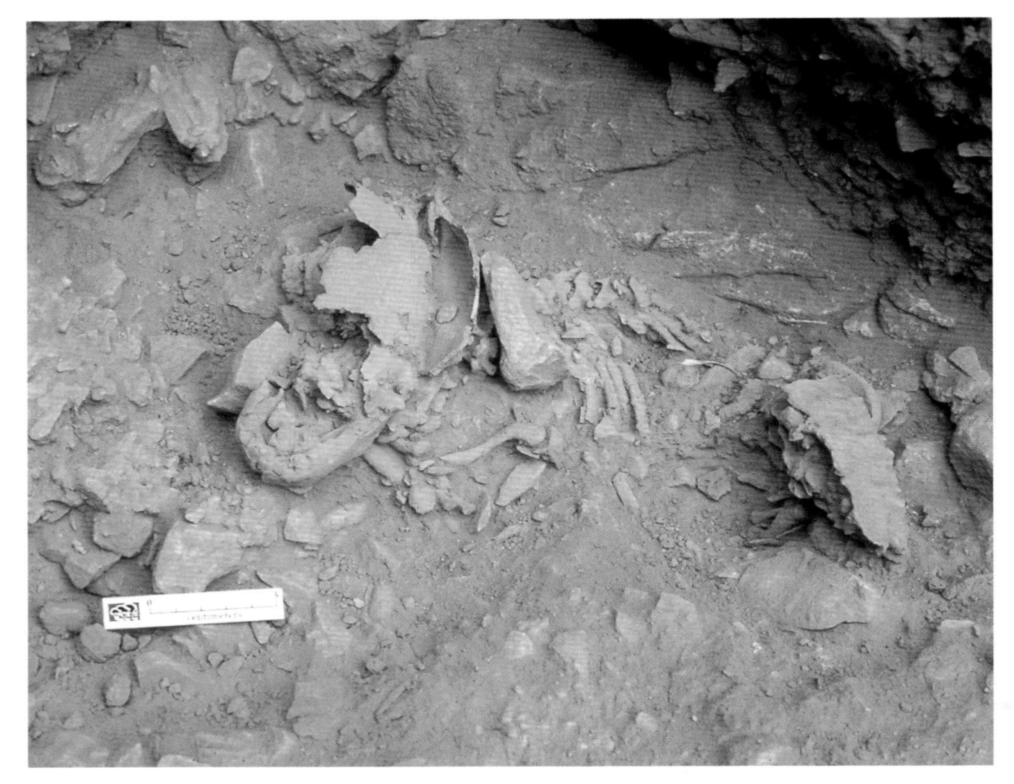

The skeleton of the Neanderthal child found at Amud by paleontologist Yoel Rak. The baby was deliberately buried in a specially dug grave.

But perhaps the most remarkable is a burial site containing two females and one male that was discovered at the easternmost extreme of the Neanderthal's range – at Shanidar in Iraq. Excavated by the US archaeologist Ralph Solecki in the late 1950s and early 1960s, this grave was covered in soil rich in the pollen of early spring wild flowers – much more than could have been blown in by the wind or carried on animals' feet. It is possible that these individuals died of starvation in late winter, when the weather was harsh and food scarce.[10] Then they were buried, and flowers

Neanderthal skeleton found at Kebara, in Israel. In addition to the fossil bones unearthed there, scientists have recovered more than 25,000 stone implements, some of which appear to have been hafted to wooden shafts and used as spears or lances. Several hearths on which Neanderthals cooked gazelle and deer meat have also been found at Kebara.

were scattered around their bodies by the rest of the tribe. Solecki was so struck by this discovery – and by the notion that Neanderthals were caring, gentle people – that he called his 1971 account of the excavations *Shanidar: The First Flower People*. "The evidence of the flowers brings the Neanderthals closer to us in spirit than we have ever before suspected," he claimed.[11] The idea that the Neanderthals were the first hippies and dispersed peace and love is rather over the top (to put it mildly). Nevertheless, it does show how far science had come in rehabilitating the species' image.[12]

However, Shanidar had other secrets to tell – stories that were every bit as revealing as the Neanderthals' care for their dead. A total of nine skeletons were eventually excavated, of which one was spectacularly injured. "This person was about

Neanderthals clearly venerated their dead. Several burials have been unearthed and have been found to contain flowers or animal skulls with which the bodies of men, women, and children were adorned.

40 when he died, and was the most injured Neanderthal that we know about," says anthropologist Erik Trinkaus. "This guy was an amazing survivor. He had lesions from head to toe. The bone around his left eye and cheek was crushed but had healed over. He probably would have been partially blind. He also had a withered arm, severe arthritis in his right knee, ankle, and big toe, and healed fractures in one foot. And yet he lived – at least for several years after suffering these injuries. He must have been helped and sustained by his family and tribe."[13] Even the most famous of all Neanderthals – whose remains were discovered in the Neander Valley in 1856 – suffered serious injury but lived on, as Ralf Schmitz of Cologne University, Germany, has found out. "He had broken his left arm," says Schmitz. "He couldn't move it normally, so he had to do everything with his right arm. He also had lots of arthritis."[14] On his own, he would have perished. Yet he survived, indicating that Neanderthals must have looked after their sick and wounded – partly as a consequence of the tight social cohesion of their lifestyles, and partly, one assumes, because these injured individuals had something to offer their fellow tribesmen, such as wisdom and advice.

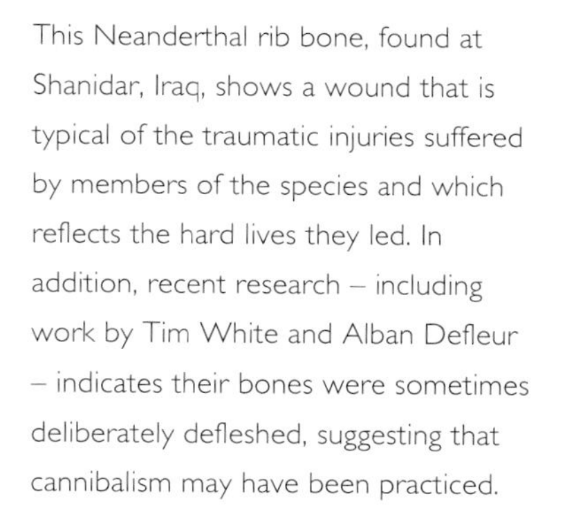

This Neanderthal rib bone, found at Shanidar, Iraq, shows a wound that is typical of the traumatic injuries suffered by members of the species and which reflects the hard lives they led. In addition, recent research – including work by Tim White and Alban Defleur – indicates their bones were sometimes deliberately defleshed, suggesting that cannibalism may have been practiced.

In addition to revealing the caring nature of Neanderthal society, the numerous injuries on their skeletons pointed to a dangerous, perhaps violent, side to their lives. Trinkaus, working with his student Tomy Berger, carried out an analysis of the bones of 17 Neanderthals – individuals who, it was revealed, had gone through a staggering total of 27 traumatic wounds. "They were mostly injured to the head and upper body, almost no lower limb injuries," says Berger.[15] "I got a statistical fit with rodeo riders. They get thrown off big animals a lot." In other words, it looked like Neanderthals were being flung around and badly hurt by the creatures that they hunted. Not for them the low-risk, careful business of stalking and spearing. They went in close for the kill and paid the consequences – although this was certainly not a mindless, headlong rush, but was probably done as a group and with a strategy. As Steven Kuhn of the University of Arizona says, "A few cooperating hunters could have exploited natural landscape features like bogs and deep stream banks that put large animals at a disadvantage. They probably killed at close range with wooden spears that perhaps had a sharp stone point."[16] These were people who had evolved a robust response to the rigors of survival, "creatures with physical prowess far beyond the aspirations of even the best Olympic athletes", as John Shea of Harvard University puts it. All the evidence indicates that they lived "a hard, violent" life, he adds. "There is the near ubiquity among Neanderthals of healed head, arm, and leg traumas. In addition, age estimates suggest that few if any Neanderthals lived beyond their 30s and 40s to a post-reproductive age."[17]

ABOVE: The blades on the right were found in French rock shelters. These Neanderthal tools display a new sophistication: they are made with a special grip on the side for the hand. The elk and wolves' teeth on the left were used as pendants.

The intriguing point is that, in all these descriptions of *Homo neanderthalensis*, we see manifestations of many of the themes discussed in earlier chapters. We had strong hints that the people of La Sima revered their dead. At Amud, Le Moustier, La Ferrassie, and other sites we see clear evidence that Neanderthals did so. In Chapter Three we saw clues that *Homo erectus* specimen KNM-ER 1808 may have survived vitamin A poisoning thanks to the care of its fellow tribe members. The badly injured people from Shanidar, Neander Valley, and other sites provide convincing evidence that Neanderthals cared for their own. And in the catalog of traumas studied by Trinkaus and his colleagues, we see fulfillment of the words of Mark Roberts and Michael Pitts, who stated that the heidelbergs and their descendants in Europe – the Neanderthals – were people who "put everything into the hunt". So many tantalizing elusive evolutionary forces were making themselves apparent in these enigmatic, contradictory people: caring and considerate, yet robust and unsubtle. No wonder mankind has had a problem with the Neanderthals.

We also know that they could build fires, although the hearths they used appear fairly simple and crude. In most cases, cave floors were simply left under layers of compressed ash. There are also ambiguous signs that they constructed shelters, and they certainly made complicated stone tools of a type known as Mousterian implements. These reveal a far greater degree of forethought and knowledge than was apparent in the Acheulian kit of *Homo erectus*. Handaxes, cleavers, and other large tools were replaced by smaller implements chipped and chiseled from more delicate flakes

NEANDERTHAL DOMAIN

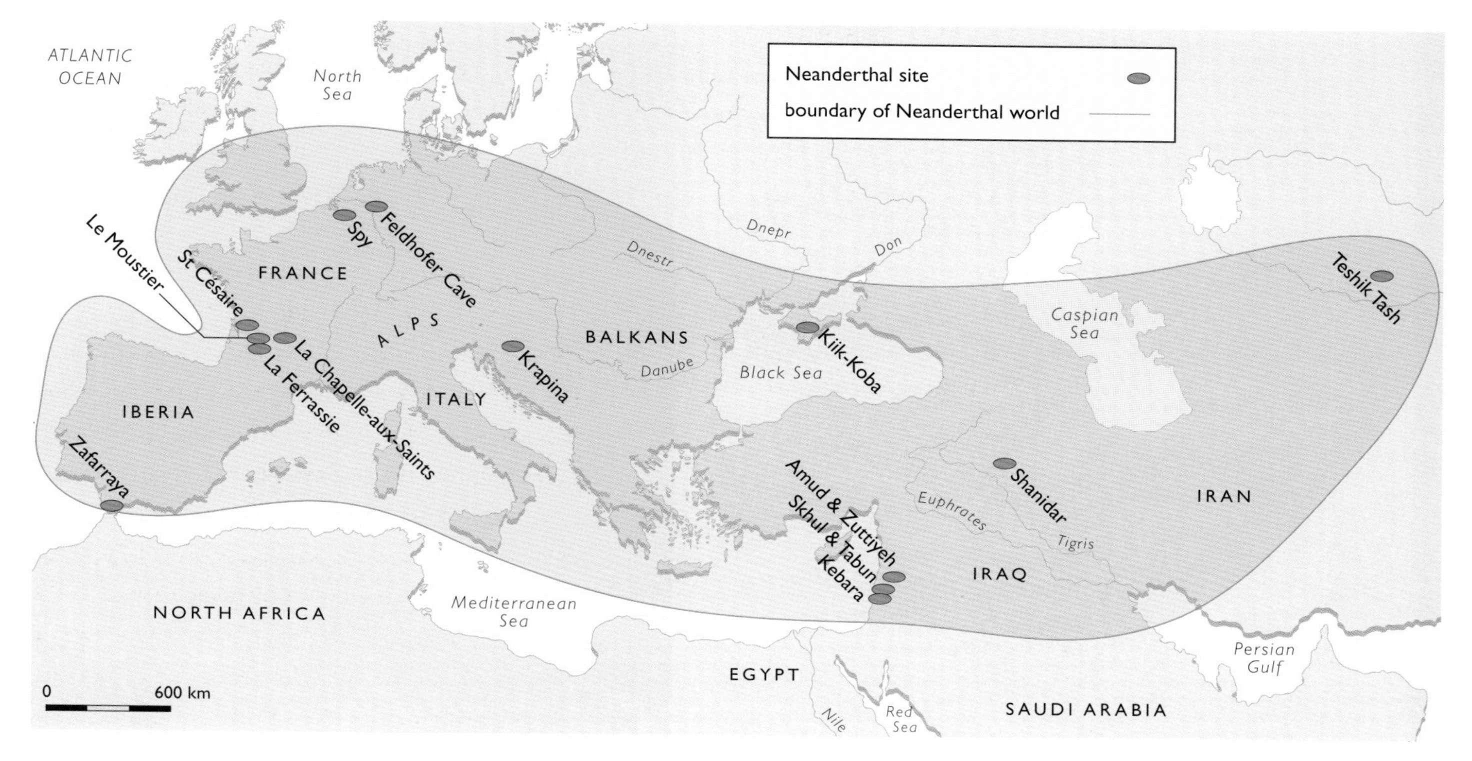

of rock. A popular device was the side scraper, which – judging from the patterns of wear – was used for scouring hide and working wood. Small stone flakes with sawtooth edges were also crafted, as well as blades that look as if they might fit onto a haft or handle to make a knife or spear.[18] And Neanderthals even made simple ornaments, such as pendants with holes for string.

In retrospect, Wells's description of these people, and the derogatory remarks of scientists such as Boule and Schaafhauson, look like grotesque propaganda. Fortunately, more recent accounts have been fairer, depicting the Neanderthals in terms of their noble simplicity, as in William Golding's novel *The Inheritors*. Compare the grim portrait painted by Wells at the beginning of this chapter with the following profile by Golding: "The mouth was wide and soft, and above the curls of the upper lip the great nostrils flared like wings. There was no bridge to the nose and the moon-shadow of the jutting brow lay just above the tip. The shadow lay most darkly in the caverns above its cheeks and the eyes were invisible in them. Above this again, the brow was a straight line fledged with hair; and above this there was nothing."[19]

The domain of the Neanderthal: remains of *Homo neanderthalensis* were first discovered at the Feldhofer Cave in Germany in 1856 and have since been found across a swathe of southern and middle Europe and also in the Middle East.

FIRE POWER

The control of fire was one of our ancestors' most powerful innovations. Not only did it provide them with the means to extend their diets by roasting otherwise toxic or indigestible foods, it also opened up a host of new techniques for obtaining vegetables or meat. For example, fire-drives made it possible to force game toward hunters, while the burning of bushes and trees stimulated the growth of new shoots and the production of nuts and other seeds.

All known hunter-gatherer tribes today cook food, although exactly when our ancestors began to control and make fires on a regular basis, it is hard to say. Remains have been found at Turkana and Swartkrans that suggest fire was used as a defense against predators as long ago as 1.3 to 1. 4 million years, in addition to signs of its use 1.6 million years ago at Koobi Fora in Kenya. This evidence is controversial, however, and many scientists doubt that humans could control fire at that time. Less contentious are the remains of putative hearths and fragments of burned animal bones have been found at sites that include Terra Amata in France, and Zhoukoudian in China. These have all been dated at between 300,000 and 400,000 years old. By contrast, no hearths were found at the Boxgrove excavations, which are about 500,000 years old, although this absence of evidence is not evidence of the absence of fires at the time.

In fact, fireplaces do not appear regularly until about 40,000 years ago, and are usually associated with the emergence of *Homo sapiens* from its African

homeland. It is therefore likely that the gift of the power of fire played a key role in our ancestors' conquest of the then freezing lands of Europe, even though these were men and women who were adapted to the heat of central Africa. As Professor David Harris of University College, London, states in

A recreation of Neanderthals gathered around a camp fire. Apart from the effects on diet that were introduced by our eating cooked food, campfires like these created a focus for the social lives of our ancestors, a place where thoughts and ideas could be exchanged and the social fabric of the tribe strengthened.

The Cambridge Encyclopedia of Human Evolution: "Making and carrying fire was essential to the occupation of areas near the ice sheets. Perhaps then, for the first time, fire began to be used to dry and smoke meat and fish, and to preserve other foods to help overcome seasonal shortages."

Poor old Neanderthal man was at last getting some decent reviews. After more than a century of vilification, derision, and dismissal as a lumbering subhuman, he was now being viewed as a creature of intellect and spirituality. Yes, he had bony brows and a heavy jaw and would have looked intimidating, but he was nowhere as monstrous as the apeman described in *The Grisly Folk*. Just reincarnate one; bathe, shave, and dress him in modern clothing; and place him in a New York subway – then "it is doubtful whether he would attract any more attention than some of its other denizens," as A.J.E. Cave, who did so much to rehabilitate the Old Man of La Chapelle, once claimed.[20] Such remarks recall the claim that the Nariokotome boy could have passed muster among humans with only "a cap to obscure his low forehead." But for Neanderthals, even headgear would have been unnecessary to hide a lack of cranial capacity. They were unreservedly human and every bit as brainy as *Homo sapiens*, with skulls in the same size range as ours – although not every scientist agrees that the Neanderthals could have passed the subway test with ease. There were those brow ridges and those mighty biceps, after all. As geneticist Steve Jones of University College, London, puts it, most people would not just shift seats if a Neanderthal, even a well-dressed one, sat beside them: "they would change trains."[21]

We can therefore see that over the decades, scientists have significantly reappraised these prehistoric people. No paleontologist would now dream of suggesting that Neanderthals had the blank, unknowing eyes of a lion. Looking at them, we would see a strange, distorted reflection of ourselves. "Neanderthals play the same role in the past that extraterrestrials play in the fiction of the future," says writer James Shreeve in his book *The Neanderthal Enigma*. "Their otherness defines our natures, highlights our failures and limitations, more than it disparages their own."[22] The point is also stressed by Fred Smith of Northern Illinois University. "Neanderthals were highly resourceful, highly intelligent creatures. They were us – only different."[23] And that is why the species plays such an important part in the story of human evolution. As hominids with a slightly different sort of intellect, physique, and behavior, they provide us with clues that help shed light on our own nature. The subtle discrepancies between them and us are revealing – and if we know what we are not, we will have a much clearer idea of what we are.[24]

The Neanderthals were like us in many ways, yet they were stronger and their brains were actually larger (although this does not mean they were more intelligent). It is an unsettling amalgam of attributes that raises compelling questions. If they were so muscular and smart, where are they now? They survived 200,000 bitter European winters and showed every sign of cultural progress. Their best artwork, their most sophisticated stone tools, and most of their graves date from their recent history – about 50,000–60,000 years ago. Then, abruptly, they disappear from the fossil record. Their last remnants, found in a few caves in southern Spain and one or two other isolated pockets, date from about 30,000 years ago. As Shreeve puts it, "The Neanderthals appear to have been outfitted to face any obstacle that the environment could put in their path. Apparently, they could not lose. And then, somehow, they lost. Just when the Neanderthals reached their most advanced expression, they suddenly vanished from the Earth."[25]

Exactly where the last Neanderthal perished we will never be able to tell, although Zafarraya, near Malaga in Spain, is as good a candidate as any. Here, in a tiny, cramped passage in a limestone cavern, researchers have discovered remains that show Neanderthals were still lingering on less than 30,000 years ago. "Southern Spain is the cul-de-sac of Europe, the very end of the continent," says one of the dig's directors, Dr. Jean-Jacques Hublin of the Musée de l'Homme in Paris. "If Neanderthals were going to hang on anywhere, it would be here."[26] At Zafarraya, Dr. Hublin's team has found a treasure trove of Stone Age detritus: flints, human and animal remains, at least one hearth, and piles of bones of the local subspecies of goat, *Capra ibex pyraneica*. And so it seems that, in this cave, a few Neanderthals survived, huddled around fires cooking goat meat, while watching the sun set over the Sierra de Alhama mountains. Then they disappeared, giving Zafarraya a unique claim to fame – as the lair of the last Neanderthal. (There are rival contenders for this title. Scientists from the former Soviet Union have reported evidence that Neanderthals were living in parts of Croatia and Crimea about 29,000 years ago, for example.)

A display at the Neanderthal Museum in Ekrath, Germany, near where the first fossil discoveries of this mysterious species were made. Once derided as a bestial throwback, scientists now recognize that Neanderthals were very like modern humans, as this rather fanciful depiction suggests.

The location of the actual last lair is not the important issue, of course. The real question is: what happened? Why are the Neanderthals not masters of the world today

WINTER'S TALE

Most people assume that the Ice Age no longer affects our planet. In fact, the Earth is still in the grip of one, and has been for the past few millions of years. At its most intense, the Ice Age has covered large parts of Europe and North America with vast sheaths of ice several miles thick. Such interludes are known as glacial periods, and the last of these loosened its grip only about 10,000 years ago. In its absence, the world now enjoys a warm climatic interval, called an interglacial period.

During the last glacial period, much of the planet's water was locked up in ice sheets and glaciers, and ocean levels dropped more than 300 ft (100m). This created land bridges across shallow seas, including one that linked Alaska and Asia, an expanse that is given the name Beringia because it stood where the Bering Straits now flow. In addition, many of the islands of Southeast Asia were joined, although Asia was never linked to Australia.

These continental connections played a critical role in the dispersal of our ancestors. During the spread of *Homo erectus*, land bridges permitted people to pass down through Malaysia, Sumatra, and into Java. Later, during the last millennia of the last glacial period,

Beringia opened up the Americas to Asian tribes (of *Homo sapiens*), who slowly migrated into the New World.

The Ice Age has also brought massive fluctuations in weather. During their recent, worst periods, they would have driven Neanderthal men and women out of northern Europe and may even have forced them as far south as the land that forms the Middle East today. It was here that the species probably had its first encounters with members of *Homo sapiens*, men and women who were destined to take over the Neanderthals' own homeland, Europe.

ABOVE: The last full glacial period began about 30,000 years ago – just as the last Neanderthals were dying out, huddled in caves in southern Europe – and continued until 13,000 years ago. The continent was covered in glaciers and ice caps (shown on the map above in white), and much of the landscape was reduced to tundra. Land areas increased (in gray on the map) as the sea level dropped, so much water having been turned to ice. Most food from this period was based on reindeer and horse meat, and although humans settled in areas quite close to the major ice sheets, such as Britain and Germany, glaciers meant these homes were frequently deserted.

LEFT: Eskimos hunting walruses. Modern humans have settled in virtually every part of the globe, no matter how inhospitable – including intensely cold polar regions. This adaptation demonstrates a flexibility of behavior that Neanderthals appear to have lacked.

A palaeontologist toils in a tiny crevice inside a cavern at Zafarraya in southern Spain. Some of the last Neanderthals are thought to have lingered on in the area after the rest of the species disappeared in northern Europe around 30,000 years ago.

instead of us? As Steve Jones asks, 'How could it have been predicted only 30,000 years ago that one moderately common primate would be among the most abundant mammals, while its genetically almost indistinguishable relative was near extinction?'[27]

Attempts to answer such questions have generated some of science's most bitter disputes, for they go directly to the heart of our sense of humanity. These issues will form the core of the last two chapters of this book. Two factors will play important roles in the story. One concerns the Cro-Magnons – the first *Homo sapiens* men and women to appear in Europe, about 40,000 years ago. They are named after the Cro-Magnon cave in France, one of the first sites at which their bones were discovered. (Cro-Magnon means 'big cliff', a reference to the limestone massif that rises above the town of Les Eyzies. It was in a cave in this cliff in 1868 that workmen constructing a railway line and station found a single grave containing four Cro-Magnon adults and a child, together with pierced shells and animal teeth that had probably been worn as necklaces.) Their relationship with the Neanderthals is crucial to our discussion. Equally important has been the emergence of a new way of studying our prehistory – not by examining the bones of the dead, but by analysing the genes of the living, and using the results to infer past population movements. This technique has shed dramatic new light on the Neanderthals and has given us new insights into our own nature. Genes, Neanderthals and mankind: it is a heady brew – as we shall now see.

CHAPTER SEVEN

MOTHER'S DAY

The artwork on the cover of *Newsweek* was unusually titillating on January 11, 1988. A young couple are seen standing beside an apple tree. They are graceful, lean, tanned – and naked. The woman is offering an apple to her partner, her breasts covered by the coiled strands of her hair. He is smiling, with his hand out, while a fat green snake watches them intently. There could be no mistaking the identity of this pair, although "with their polished copper skin and taut physiques, they seemed more like the products of an exclusive health spa than the residents of an eternal paradise," says the writer James Shreeve.[1]

Adam and Eve were making headlines again, although it was only the latter who had actually made the journalistic grade, albeit in a big way. She had achieved her new-found fame thanks to a pioneering research project carried out by Allan Wilson, Rebecca Cann, and Mark Stoneking at the University of California, Berkeley. And what this team had to say was truly astonishing. They had concluded that every person on Earth could trace his or her pedigree back to one woman, an African, who had lived a mere 200,000 years ago. New Guinean tribesman, Parisian bartender, American teacher, and Polynesian farmer: they may seem disparate, improbable relatives, but all are linked through this woman, the Berkeley scientists said.

It was a dramatic announcement and it marked the day that molecular biology began to make a serious impact on the investigation of human evolution – genes had joined bones and stones as appropriate subjects for evolutionary studies. Of course, our genes' principal role is to control the growth and development of the living cells of our bodies, but they can also act as emissaries from the past, providing us with information about lost generations. Their value to prehistorians is neatly summed up by the British anthropologist Jonathan Kingdon. "Fossil bones and footsteps and ruined homes are the solid facts of history, but the surest hints, the most enduring signs, lie in those minuscule genes. For a moment we protect them with our lives, then like relay runners with a baton, we pass them on to be carried by our descendants. There is a poetry in genetics which is more difficult to discern in broken bones, and genes are the only unbroken living thread that weaves back and forth through all those boneyards."[2]

In the 1980s, research by geneticists such as Luca Cavalli-Sforza of Stanford University, California, had already begun to trace movements of prehistoric populations by studying variations in genes extracted from modern people of different races. However, it was the Berkeley investigation that caught the public's imagination, so stark and bold were the team's conclusions. Wilson and his colleagues had taken a piece of tissue from the placenta of women from various ethnic groups and, from these samples, isolated a special type of genetic material called mitochondrial DNA. Mitochondrial DNA comes from tiny structures called mitochondria, which are found inside nearly all the human body's cells. The DNA in mitochondria is useful for studying human evolution for two important reasons. First, nearly all mitochondrial DNA is passed from one generation to the next by mothers only, so written within it is a history of the world's women. And second, unlike the DNA inside the nuclei of our cells – where our genes are stored – mitochondrial DNA mutates very rapidly. Over the generations, these mutations build up at a steady rate. By comparing the mutations in different ethnic races, scientists can work out a family tree of modern humanity and estimate how long ago the various races split from a common ancestor. The Berkeley team produced a tree that had two main branches: one made up of Africans and another made up of the rest of the world's people. At the tree's base was the common ancestor of all humanity, who lived in Africa about 200,000 years ago, the Berkeley team calculated. This was the founder of *Homo sapiens* and it first appeared in Africa a mere 200 millennia ago. Given the antiquity of the human lineage, which we have traced back 5 million years in this book, such a figure was astonishingly recent.

Not surprisingly, the investigators' conclusions, outlined in the journal *Nature* in January 1987, made headlines around the world, particularly since Wilson argued that he could trace humanity's roots not just to a single group or tribe but to an individual woman – the first female of our species, or "African Eve,"[3] as

Back in the news: Adam and Eve as depicted by *Newsweek*. The magazine made a cover story of the report that scientists had traced human antecedents back to a single woman who had lived in Africa 200,000 years ago.

she was dubbed. "Eve was more likely a dark-haired, black-skinned woman, roaming a hot savanna in search of food," said *Newsweek*. "She was as muscular as Martina Navratilova, maybe stronger; she might have torn animals apart with her hands, although she probably preferred to use stone tools. She was not the only woman on Earth, nor necessarily the most attractive or maternal. She was simply the most fruitful, if that is measured by success in propagating a certain set of genes. Hers seem to be in all humans living today: 5 billion blood relatives. She was, by one rough estimate, your 10,000th great-grandmother."[4]

The timing of her discovery could not have been better, for paleontologists were then fiercely divided over the questions that we raised at the end of the previous chapter. What had happened to the Neanderthals? What role did *Homo sapiens* play in their demise? And where did modern humans come from? The debate was split between two camps: the "Out of Africa" theorists and the "multiregionalists." A brief examination of their views, starting with the latter group, will reveal why the Berkeley paper had such a dramatic impact.

For several decades, one group of anthropologists had argued that Neanderthals had disappeared from Europe's fossil record for the simple reason that they had evolved into another species: modern humans. The idea of this transformation came in the wake of scientists' reappraisal of the Neanderthals. Having been so damned by Boule, Neanderthals were slowly rehabilitated as the 20th century progressed. One of the most important figures in this reassessment was the Jewish, German-born scientist Franz Weidenreich. He had made his name by heading the institute that organized expeditions that had discovered *Homo erectus* remains in China, before settling in New York when he was forced to flee Nazi Germany. Weidenreich believed each of the world's inhabited regions had its own local lines of human evolution "which all proceeded in the same general direction with mankind of today as their goal." In particular, he argued that Neanderthal man had given birth to *Homo sapiens*. You could trace a line from *erectus*, which evolved into Neanderthals, which, in turn, were slowly transformed into *Homo sapiens* in Europe, he said. Similarly, a lineage could be followed from Chinese *Homo erectus* to modern oriental humans, while another led from early to late *Homo erectus* in Java and on to present-day native Australians.

In other words, when *Homo erectus* began its great diaspora more than a million years ago, it created populations throughout the Old World that have been evolving there ever since. Pygmies, Australian aborigines, Inuits, and all the diverse peoples of the world today are the end result of this very ancient separation of mankind, and their features are leftovers from that ancestry. For example, when *Homo neanderthalensis* eventually turned into modern Europeans, one by-product was the big nose of the modern European. Similarly, flat faces in orientals and flat foreheads in Australian aborigines can be traced back to *Homo erectus*.[5] Weidenreich's ideas rehabilitated the poor old Neanderthal by placing him firmly in the role of immediate ancestor of modern Europeans, but the theory also suggested the existence of deep divides between modern humanity's constituent races.

This idea – that *Homo sapiens* achieved its current state through a process of a million years' continuous evolution across the Old World – was taken up by Loring Brace, a fierce champion of the Neanderthal cause, and subsequently by his colleague at Michigan University, Milford Wolpoff, and by Alan Thorne of the Australian National University. They too argued that *Homo erectus* had been evolving for over a million years in the valleys, plains, and mountains of Africa, Asia, and Europe, but they stressed that genes flowing between populations – "the results of an ancient history of population connections and mate exchanges" – had mitigated the effects of geographic separation on the races. Caucasians, for example, evolved from Neanderthals in Europe – with an input of genes from the rest of the Old World. This, then, is multiregionalism, a version of Weidenreich's account of human evolution refined to include the flow of genes between populations, a drift that would have offset, to some degree, the development of deep divides between races.

However, by the 1980s this idea was under attack by several sets of scientists who could see no connection between Neanderthals and modern Europeans, nor between Peking man and the Chinese of today. These researchers spoke not of continuity but of replacement. They believed that, about 400,000 years ago, our ancestors split into two populations, a division that was triggered by the spread of the Sahara Desert. The northern group became Neanderthals, while the southern one evolved into *Homo sapiens*.

It was around the time of this debate that Chris Stringer – who was to become a particularly committed advocate of the Out of Africa theory – began one of paleontology's oddest odysseys. During the summer of 1971, the young postgraduate toured the museums of Europe – in an old Morris Minor with a tent for accommodation – where the main Neanderthal collections had been gathering dust for decades. He began making systematic measurements of the bones, and returned

THE WAY WE WERE

Human beings are very different creatures today from those first members of the hominid line who moved out the trees onto the savanna floor of Africa 5 million years ago. A cascade of different evolutionary forces has shaped us in the intervening period: for example, climatic shifts altered the availability of food. Those of our lineage who adopted a flexible, omnivorous response to those changes were better able to get the nutrition and energy that permitted their brains to expand.

That brain growth in turn permitted all sorts of shifts in behavior. For example, we invented fire, and tools to pound and mash food, and that has had a powerful impact on our physiques. We no longer need mighty sets of dentures to grind and crush our food, and our jaws and teeth have shrunk as a result. The human face no longer has a muzzle, and is far shorter compared to any of our predecessors.

We have also lost our pelts of hair, apart from patches on our head (which protect us from the sun's rays), our armpits and crotches (which may be involved in the dissemination of glandular chemicals called pheromones), and on the faces of men (where they may have been intended to act as visual devices to impress enemies and females). In short, human beings still bear the anatomical stigmata of our recent evolution.

On the right, the principal anatomical changes that have taken place over the eons, and the advantages they confer on *Homo sapiens*, are outlined.

BRAINS

The brain of *Homo sapiens* is distinctive not just for its size but also for its shape. In particular, the forehead of modern humans is remarkably prominent. It is as if our crania have been squeezed at the front and back as our brain became larger and rounder. By contrast our brow ridges are puny compared with those of any of our predecessors.

ARMS

While humans can completely straighten their arms, only apes can lock their elbows to prevent dislocation, crucial for four-legged locomotion.

HANDS

The similarity between a human hand and a chimpanzee hand is striking: right down to the fingerprints that distinguish individuals of both species. One special attribute that distinguishes our ape lineage is the ability to rotate our thumbs to grasp objects. This feature first appeared about 18 million years ago.

FEET

The foot of *Homo sapiens* also plays a critical role in walking. The heel is designed to absorb the energy of impact when the foot strikes the ground. The body's weight then moves forward to be taken up by the toes, which launch the next step.

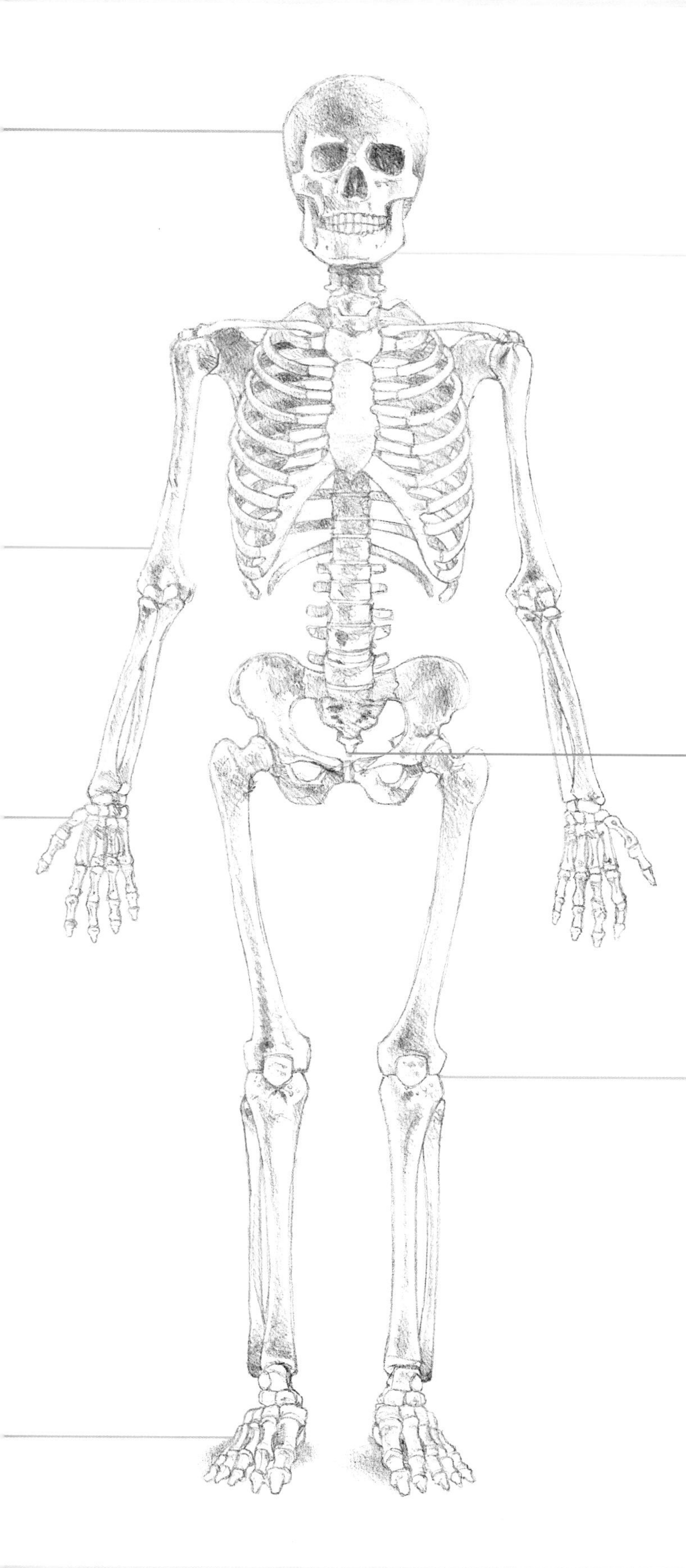

JAWS

Our jaws have shrunk. No longer required to grind vegetation or uncooked meat, our dentures have declined in size over the millennia. As a result, human faces are flat, while we have grown chins to provide external strengthening for the jaw. These facial proportions, including the shape of the jaw – short and deep – have contributed to our ability to produce human language.

TAILS

In common with all other apes, human beings do not have tails. These disappeared from our ape lineage about 25 million years ago.

KNEES

The human knee is the key to our upright gait. It can lock to create a straight leg, thus saving energy that would otherwise be needed to support our bodies. Importantly, it is also tucked under the pelvis, thus saving us from waddling and wasting energy as we move.

home five months later (having lost most of his possessions to car thieves and his tent to a Prague thunderstorm) carrying a notebook of precious data on the anatomy of *Homo neanderthalensis* and the Cro-Magnons, which were introduced at the end of the previous chapter. A statistical analysis of his data revealed no sign that Neanderthals were evolving into modern humans. According to their skeletons, these were two very different sets of people. Neanderthals had simply disappeared from the fossil record about 10,000 years after the appearance of the Cro-Magnons. "The latter had simply replaced the former," says Stringer, who is now a paleontologist at the Natural History Museum in London.[6]

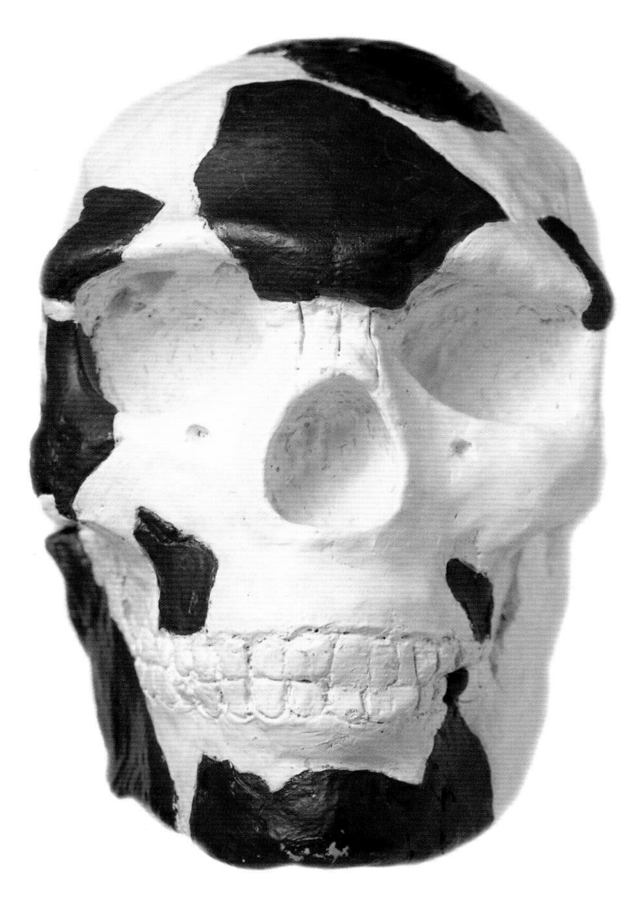

Discovered by Richard Leakey, the skull of Kibish man was found to have many of the features of modern humans, yet was at least 130,000 years old.

But where had the Cro-Magnons come from? A strong hint was provided by anthropologist Erik Trinkaus. As we saw in the case of the Nariokotome boy (see Chapter Three), people adapted to hot, dry climates tend to be tall and slim, while those adapted to cold climates tend to be shorter and stockier. These different body shapes are revealed by comparing the lengths of the shin and thigh (or forearm and upper arm). The longer the former to the latter, the hotter the climate of origin. When Trinkaus fed measurements of Neanderthal bones into this "limb thermometer," the result suggested that they were adapted to an average temperature of about 32°F – very like Ice Age Europe. But with the Cro-Magnons he got a very different figure, one that indicated their homeland was 70°F, suggesting they came not from freezing, glacier-capped Europe but from the Tropics.

The results were supported by other fossil finds, one of the most important of these being the handiwork of one of the ever-industrious Leakey clan, in this case Richard. On a 1967 expedition to the Omo-Kibish region of Ethiopia, he found a partial skull and skeleton (along with a second skull), which were dated at about 130,000 years old. At the time, this seemed an unremarkable date for a supposedly primitive hominid. Then scientists looked in more detail at the remains of Kibish man, as the finds came to be known, and realized that he was actually a modern human. The Omo I skull had a short, broad face and high forehead, the lower jaw had a chin, and the pair of

Border Cave in South Africa, where the 90,000-year-old remains of four modern-looking human beings – including a tiny infant in a grave – were discovered by scientists.

worn teeth that had survived with the bones appeared modern in both size and shape.[7] It was then that the importance of the find really began to be appreciated, as Leakey recalls in his book *One Life*. "Geological investigations and dating have shown that the two skulls are about 130,000 years old, yet despite their antiquity, they are both clearly identifiable as *Homo sapiens*, our own species. At the time of their discovery, scientists generally believed that our species had only emerged in the last 60,000 years and many considered the famous Neanderthal man to be the immediate precursor to ourselves. The Omo fossils thus provided important evidence that this was not so."[8]

ABOVE: A recreation of early modern man shown fishing with a harpoon or simple sharpened stick. Harpoons such as that shown opposite may also have been used.

BELOW: These stone implements, found at Klasies River in South Africa, along with the remains of *sapiens*-like hominids, were dated as being more than 120,000 years old.

This evidence was supplemented by a series of important discoveries in Africa. At Border Cave in South Africa, paleontologists found the remains of four modern-looking human beings, including a tiny infant who had been placed in a grave. All of them were at least 90,000 years old. At Klasies River in South Africa, *sapiens*-like hominids dated as being more than 120,000 years old were uncovered, along with some surprisingly advanced stone implements. And at Katanda in the Democratic Republic of the Congo, John Yellen of the National Science Foundation in Washington and Alison Brooks of George Washington University found superbly carved, 90,000-year-old bone harpoons and knives, the sophistication of which was a match for the handiwork of the Cro-Magnons – who had been thought to be the first to develop intricate carving skills 50,000 years later. It was as if "a prototype Pontiac car had been found in the attic of Leonardo da Vinci," as James Shreeve put it.[9]

The conclusion is straightforward, says Stringer. *Homo sapiens* is an African species, while Neanderthals "are more like respected evolutionary siblings, or even cousins." According to the Out of Africa theory (or the Noah's Ark hypothesis, as it is also known), *Homo erectus* emerged from Africa and started to colonize the world's remote corners about a million years ago, later evolving into Neanderthals (and possibly other unknown species in obscure parts of the Old World). At this point, both theories – multiregionalism and replacement – agree. Then came the twist: about 150,000–100,000 years ago a second wave of hominid upstarts – *Homo sapiens* – left Africa and replaced all those other lineages. By about 60,000 years ago they had reached Indonesia and Australia. Then, about 40,000 years ago, modern humans began to infiltrate the European domain of the Neanderthal. By 30,000 years ago *Homo neanderthalensis* was extinct. Despite having evolved in the icy rigors of this glacier-encrusted continent, Neanderthals were substituted by a race of hominids that had evolved in hot, tropical climates. Of course, some intermingling of *Homo sapiens* with Neanderthals may have occurred, but not much. Replacement is the key word in the Out of Africa theory. As Stringer says, "We are all Africans under the skin."[10]

The idea is strongly dismissed by the multiregionalists. Deriding the theory as "the great Leap Backward," Loring Brace claims: "It was the fate of the Neanderthal to give rise to modern man, and, as frequently happened to members of the older generation in this changing world, to have been perceived in caricature, rejected and disavowed by their own offspring, *Homo sapiens*."[11] Wolpoff is equally withering and depicts the Out of Africa theory, with its emphasis on the replacement of human

This intricately carved bone harpoon, found at Katanda in the Congo, was found to be 90,000 years old. Carving of such a sophisticated nature was not thought to have developed until the arrival of the Cro-Magnons some 50,000 years later.

DNA:
LIFE'S MOTHER TONGUE

The discovery of the structure of DNA, by Francis Crick and James Watson in 1953, remains one of the greatest biological breakthroughs of modern times. The two young researchers showed that deoxyribonucleic acid (DNA), the molecule from which our genes are made, was shaped like a spiral staircase; in other words, they found that it formed a tiny double helix.

In each cell in our bodies, we have about 100,000 genes, and these are all made out of double helix-shaped molecules of DNA that control the development of our muscles, organs, and brains. The discovery of how DNA dictates this growth became possible with the uncovering of its structure and has proved to be a major stimulus in creating new medicines and understanding illness.

But DNA technology has also provided the study of human evolution with a powerful new tool. For example, it allows scientists to compare people's DNA to detect minute variations between them. These studies generate crucial information about the relationships between the world's races.

In doing this, researchers exploit a special type of genetic material called mitochondrial DNA. Unlike the nuclear variety (described above), which we inherit in equal amounts from our parents, DNA from our mitochondria – tiny bodies that supply energy for our cells – is inherited only through the maternal line. Everyone has the same mitochondrial DNA as their mother, as did she, right back into the mists of time.

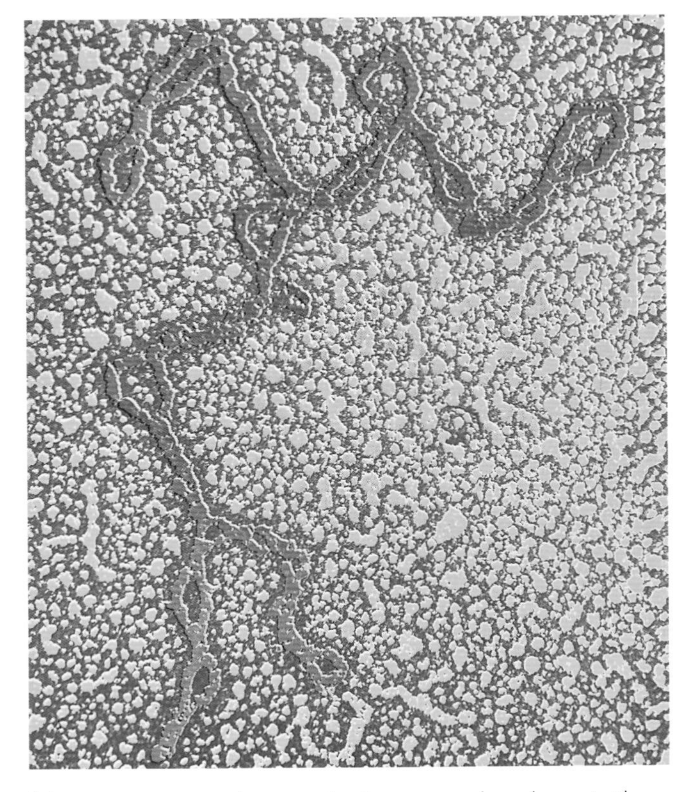

However, over the centuries, occasional mutations appear in mitochondrial DNA, and by analyzing patterns of these changes among populations, scientists can not only estimate the genetic differences between races but can also calculate when they shared a common ancestor.

In addition, a technique called polymerase chain reaction (PCR) allows scientists to make millionfold copies of bits of DNA, thus greatly improving their ability to detect genes in ancient bones and skulls. This technology has been used to study the genetic makeup of Neanderthals, and to determine how close they were biologically to *Homo sapiens*.

ABOVE LEFT: A color-enhanced transmission electron micrograph of DNA, taken from a fragmented mitochondrion.
ABOVE: A researcher examines gel containing DNA fragments. The DNA pieces have been stained with ethidium bromide and can be seen under ultraviolet light as fluorescent violet bands. An electric current is then passed through the suspension and the fragments of DNA are separated out according to size. This technique is routinely used for DNA fingerprint analysis and other genetic studies.

species, as a Stone Age holocaust. "It amounts to killer Africans with Rambo-like technology sweeping across the world and obliterating everybody they meet," he says. "Hardly my idea of universal brotherhood."[12] Not so much a Noah's Ark, more a form of species-cleansing, in other words.

For his part, Stringer responds that at no point had he "suggested a violent replacement of Neanderthals by Cro-Magnons. The two may have coexisted peacefully for several thousand years before the former group died out because it could not compete, economically, with the latter."[13] And so the arguments intensified, with multiregionalists claiming that their adversaries supported ideas that mankind was inherently violent, while the Out of Africa camp pointed out that their opponents' theories implied that humanity was still deeply divided along racial lines. This was bitter stuff – even by the normal standards of anthropologists, who, according to *Newsweek*, "have few rivals at scholarly sniping."[14]

Then Eve arrived, a mitochondrial bombshell whose existence provided the Out of Africa supporters with unexpected intellectual succor. Here, after all, was a team of researchers who had nothing to do with paleontology or anthropology, who worked in cool, high-tech American campuses instead of the parched African bush, but who could nevertheless see clear signs of a recent African origin in the distribution and genetic makeup of mankind today. Even the timescale was right. Wilson and his team were suggesting that, about 200,000 years ago, a small population of hominids evolved into *Homo sapiens*. Then they began to expand, emerging from Africa about 100,000 years ago to spread slowly around the globe. Of course, Eve was not the actual mother of humanity, they insisted; rather, she was representative of a population bottleneck through which *Homo sapiens* was passing. "She wasn't the literal mother of us all, just the female from whom all our mitochondrial DNA derives," said Wilson.

Wilson's study seemed to be too good to be true, at least for the Out of Africa camp. And, of course, it was. Its conclusions were carefully scrutinized by geneticists worldwide, and flaws and criticisms began to emerge. For example, only two of the 20 "Africans" included in the study had actually been born there. So the Berkeley team repeated their work using a geographically wider mix of samples and again produced a tree that put mankind's birthplace firmly, and recently, in Africa. It seemed settled, until Alan Templeton, a geneticist at Washington University in St Louis, denounced the Berkeley team's computing techniques.[15] "As a result of analyzing just one run, they fooled themselves into thinking they had a well-resolved evolutionary tree," said

Templeton. In fact, there were thousands of equally good but different trees that could be made from the Berkeley data, he pointed out. Other geneticists subsequently agreed, but, unlike Templeton, did not conclude that this killed off Eve. For example, Maryellen Ruvolo of Harvard University tested Wilson's results and did indeed find that they could produce thousands of different trees, but nearly all these mitochondrial bushes were only trivially different from each other. "In fact, we found three groups of trees – although there may be more if one searched further," says Ruvolo. "Two have their roots in Africa, while the third's origins are unclear. So there is still evidence of an African origin, but it is not proof."[16]

But Templeton's critique was not the only attack that the Eve hypothesis was to face. Wilson and his team had made two key assumptions: that mitochondrial DNA is inherited only on the female side, and that it accumulates mutations at a fixed rate. But research published in March 1999 by the distinguished biologist John Maynard Smith of Sussex University, along with a team from Otago University in New Zealand led by Dr Erica Hagelberg, suggested that mitochondrial DNA can be passed on from males – admittedly in small amounts, but enough to disrupt the calibration of the mitochondrial clock that Wilson had used to date Eve as being 200,000 years old. The effect would be to increase her age, perhaps to about 400,000 years old – still not enough to invalidate the Out of Africa theory, but sufficient to suggest that great care should be taken in interpreting molecular biological studies of human populations.

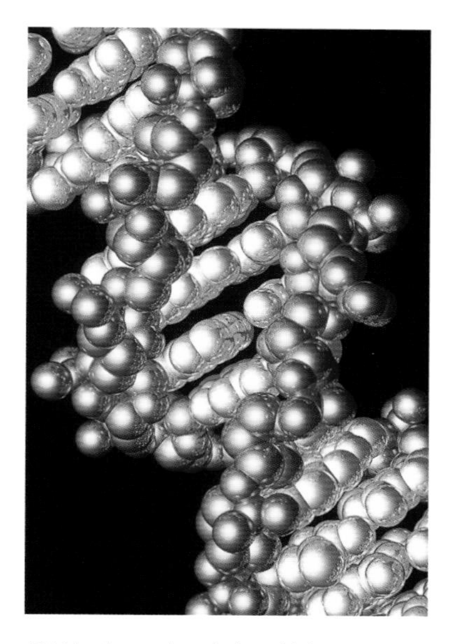

DNA, the molecule in which our genetic heritage is coded. Deciphering the messages contained in its spiral structure is providing scientists with a new method for uncovering the secrets of human prehistory.

In any case, more genetic support for the Out of Africa theory was provided by Luca Cavalli-Sforza in his massive work *The History and Geography of Human Genes*, coauthored with Paolo Menozzi and Alberto Piazza.[17] More than 70,000 frequencies of various gene types in nearly 7,000 human population types were included in this study, which showed that the division between Africans and non-Africans was the first major separation to occur in *Homo sapiens*, and that it took place about 100,000 years ago. Asians and Australians then separated about 50,000 years ago, and Europeans and Asians about 30,000 years ago. In addition, Dr. Michael Hammer of the University of Arizona has examined a stretch of the Y-chromosome – the sex chromosome found only in males – and concluded that the world's men have a common ancestor, a "nuclear Adam," who lived about 188,000 years ago. Both studies are, therefore, consistent with the Out of Africa theory.

And then, in 1997, researchers made an announcement that was every bit as dramatic as the discovery of Eve. Led by Svante Pääbo of Munich University, scientists isolated scraps of DNA from a Neanderthal bone – and not from just any old fossil, but from the very skeleton that triggered, in 1856, all that furor and fuss about our "debased" predecessor, the man from Neander Valley.

Bone cells have nuclei containing DNA, like other cells. However, it is extremely difficult to extract such material from fossilized skeletons. Geneticists had been trying for years to obtain Neanderthal DNA, but had failed because DNA's delicate spiral columns degrade relatively quickly. In good conditions they can remain intact, in patches, for a few tens of thousands of years – enough to think of analyzing Neanderthal DNA, but not much more (and certainly not enough to clone a 65-million-year-old dinosaur, as depicted in *Jurassic Park*). Then Pääbo and his team delicately sandblasted 0.12 oz (3.5 g) of DNA from the upper armbone of the man from the Neander Valley and struck genetic gold. The cold of the valley cave seems to have slowed his DNA's disintegration. The fact that his bones were varnished – a practice intended to help preserve remains but which is now discontinued – may also have helped. In addition, he was probably one of the last of his lineage, so his DNA had less time to disintegrate. Pääbo used a technique known as DNA amplification to make millionfold copies of each scrap of the genetic material – enough to let him determine its exact chemical structure. He managed to recover a snippet of DNA that was 379 base-pairs long. (A base-pair is the basic unit of DNA. Think of each one as a step on the spiral staircase that makes up the double-helix structure of DNA.) Compared with the 3 billion base-pairs that exist in human DNA, this was not a staggeringly long piece. Nevertheless, it was enough to make a critical comparison with *Homo sapiens*, because, when Pääbo looked at his results, he found he had obtained DNA quite unlike that of any human being alive today. There were 27 differences between *Homo sapiens* and Neanderthal DNA along this 379 base-pair section – a far greater variation than ever occurs between different human groups today. The entire population of the world differs by only eight base-pairs on this DNA strip. "The DNA was quite like a human's, but not exactly," says Dr. Thomas Lindahl of the Imperial Cancer Research Fund.[18]

Of course, some form of contamination could have skewed the results, but this seems unlikely given that Pääbo's work was duplicated by Mark Stoneking, who was now based at Pennsylvania State University. Stoneking took a different sample from the same bone, amplified it, and isolated a sequence that exactly matched a large chunk of the Munich DNA. In short, scientists had cloned a piece of Neanderthal DNA and found it was substantially different from ours. Only a species that began evolving separately from our lineage about 600,000 years ago could produce such variation, they concluded.[19] The conclusion was a blow for the multiregionalists, who believed Neanderthals were our direct ancestors rather than a separate lineage of ancient origin.

So where does this leave the two sides in our paleontological battle? Have the multiregionalists succumbed, or have they successfully fought off the predations of the Out of Africa protagonists? Has one side won and the other lost? The answer to the last question is simply no. No knockout blow has yet been delivered. However, of the two camps, the Out of Africa side looks by far the stronger, and most scientists now assume that modern humanity is made up of people of recent African origin. Despite our apparent differences, mankind is a surprisingly homogenous species – a point revealed in studies that show there is more genetic diversity in a social group of chimpanzees or a clan of gorillas in a single forest than there is throughout the entire human population.[20] As David Woodruff of the University of California, San Diego, puts it, "From a chimpanzee's point of view, we all look like Dolly." In other words, we are virtually clones, so similar is our DNA. We display remarkable geographic diversity, but astonishing genetic unity – because the world's different peoples have such a recent common ancestry. This has important implications that we shall discuss in the next chapter. The crucial point is that the notion that we are all recent African arrivistes is, at present, in the ascendancy, partly because it commands the more convincing fossil and genetic evidence, and partly because it is intellectually more satisfying, a point emphasized by Stephen Jay Gould, who believes that "multiregionalism will probably be remembered as the last post of the linear view." As he points out, its proponents argue that *Homo erectus* started to evolve in Asia, Africa, and Europe in parallel (abetted by a low level of migration) toward *Homo sapiens*. "Such an idea represents linearity with a vengeance as all subgroups within a single species move onward (and brainward) in the same optimal direction." By contrast, adds Gould, in the Out of Africa alternative, *Homo sapiens* arises as a branch from the bush of hominid evolution and "not as a terminus to a universal trend."[21] And after all, this latter model is the one we use to interpret the evolution of all other species, he adds. "We have no multiregional theory for the origin of rats or pigeons, two species that match our success and geographic spread. No one envisions proto-rats on all continents evolving together toward improved ratitude. Rather we assume that *Rattus rattus* and *Columbia livia* initially arose in a single place, as an entity or isolated population, and then spread out, eventually to cover the globe."[22] And thus it should be with *Homo sapiens*.

For the rest of this book we shall adopt Gould's stance and assume, as do most modern scientists, that the arrival of Cro-Magnons in Europe and the disappearance of the Neanderthals shortly afterward were connected – and in one quite specific way: that the former was the cause of the latter. But just how total was that replacement? Did *Homo sapiens* simply take over, and did *Homo neanderthalensis* die without issue? Was there interbreeding? These are good questions, as yet without clear answers. It should be noted, however, that Out of Africa proponents did not suggest when they first outlined their theory that there had been total substitution of one population for another. Some mating between Neanderthal and modern human was always considered a likelihood, although most protagonists argued that it was not common. Neanderthals may have had some input into modern gene pools, but not much. The real surprise has been that virtually no input has been detected at all. Our takeover involved virtually no fraternizing with the locals, it would seem. Certainly, the genetic study carried out by Pääbo indicates quite sharp differences between our DNA and that of the Neanderthals.

The skeleton of a child, unearthed in 1999 by Portuguese researcher Joao Zilhao. The body has features of both modern human and Neanderthals, according to some scientists, and indicates that the two species could have interbred.

The fossil record is similarly unsupportive of the idea – except for one key discovery: a child's skeleton found in Portugal in 1999 by Joao Zilhao of the Instituto Nacional de Arqueologia, Lisbon. Originally thought to be the skeleton of a modern human, it was subsequently studied by Erik Trinkaus, who announced that it showed distinctive Neanderthal features, as well as features typical of *Homo sapiens*. "This is the first unequivocal proof that Neanderthals and humans interbred," says Trinkaus. "It shows that Neanderthals and humans lived together and loved together." This raises the possibility that we carry Neanderthal genes among our own; Trinkaus is circumspect, although not dismissive. "It's possible. I don't think we'll ever find out, for I doubt we'll ever get anything that contains the full Neanderthal genome. But

Neanderthal and human genes were mixing 40,000 years ago."[23] Other scientists disagree, however. "Nothing resembling Neanderthal DNA has ever shown up in all the analyses of human genetic material," says Cambridge University archaeologist Paul Mellars. "This argues strongly that the amount of interbreeding between Neanderthals and humans was very small – if, indeed, it happened at all."[24] The Portuguese skeleton has yet to be studied by other scientists, so the jury remains out on this specific example, although it should be stressed again that some interbreeding is not ruled out by Out of Africa theorists.

This leaves us with one critical question, and it is probably the most important one in this book, the issue to which our entire discussion has been leading: why did *Homo sapiens* triumph? Neanderthals were both brainy and strong, yet they perished. Why? What special attribute did our species possess that allowed it to triumph at the expense of other hominids? How could men and women adapted to tropical climes have succeeded in conquering all the intemperate regions of the world? Answers to these questions will complete the story of our evolution and will form the core of this book's final chapter. In it, we will concentrate on what it means to be a member of *Homo sapiens*, although occasionally we will have to make comparisons between ourselves and other hominids, in particular the Neanderthals – because, as we have said, if we know what we are not, we will have a much clearer idea of what we are. We should take care in our interpretations, however. In explaining why Neanderthals failed and why we succeeded, we run the perennial risk of implying that our current global domination was decreed from the start, a trap that awaits all who explore human evolution. Good fortune, not predestined greatness, has been the leitmotif of the story of *Homo sapiens*. The following pages explain why we got lucky.

CHAPTER EIGHT

HOME ALONE

On December 18, 1994, three potholers were scraping through the tight, twisted corridors of a cavern in the Ardeche Gorge of southeastern France. It was a regular Sunday jaunt for the trio; scrabbling through a pitch-black labyrinth was their idea of weekend jollity. At one point, Jean-Marie Chauvet, Eliette Brunel-Deschamps, and Christian Hillaire had to squeeze through a particularly convoluted corridor, and felt an unexpected draft of cold air gust through a pile of rocks. The cavers dug away the boulders, cleared a tight passageway, and revealed a shaft that plunged down into the darkness. They used rope-ladders to descend into a cavern 30 ft (9 m) below, and shone their lights at the glistening walls. What they saw stopped them in their tracks, for there, in the beams of their helmet lamps, was a breathtaking vista: gigantic columns of white and orange calcite, draperies of sparkling minerals, and bear bones scattered on the ground. Then Eliette spotted an even greater wonder: an image of a mammoth. The potholers looked closer and saw that the cavern walls were bristling with engravings and paintings in bright red ochre and black charcoal, a bestiary of astounding vividness that included more than 300 animals, among them rhinoceroses, lions, buffaloes, panthers, and deer tumbling in profusion across the cavern walls. The creatures were depicted with originality and flair; Indiana Jones never discovered anything half as exciting. "We had a date with prehistory that day," Chauvet later recalled.[1]

Cave painting of a horse, made by a Cro-Magnon artist in Grotte Combe d'Arc cavern at Chauvet in France.

The cavern was named the Grotte Chauvet after Jean-Marie, and is now regarded as one of the world's greatest galleries of parietal art (*paries* being a Latin word for wall).[2] And it is a true anthropological wonder, one that reveals just how stunning the impact of its masterminds – the Cro-Magnons – must have been on the world. Many of the creatures on Chauvet's walls were painted in perspective, an effect the artists heightened by carefully scraping away parts of the wall to enhance contours. The paint was applied delicately, both with fingers and tools, to create shading and relief. "The people who did this were great artists," pronounced France's leading cave art expert, Jean Clottes.[3]

But there was more to the potholers' discovery than the mere addition of a new set of paintings to the world's existing collection of Cro-Magnon galleries. Their cave art was already well known, and included several spectacular sites at Lascaux, Altamira, and Les Eyzies (including those that had enticed the young Mary Leakey to a career in archaeology, as we saw in Chapter One). But, of course, "Cro-Magnons are us – by both bodily anatomy and parietal art – not some stooped and grunting distant ancestor," as Stephen Jay Gould puts it.[4] We should not be astounded that these people could create works that display perspective and originality.

Yet we are surprised. Temporal chauvinism is hard to discard, as we have already seen. We are inured to the idea that we are the pinnacles of a linear evolutionary ascent. And so it is with artistic ability. Until the cavern at Chauvet was uncovered, most experts in prehistoric paintings believed they could see conspicuous signs of improvement in the quality of Cro-Magnon art, running from the earliest works, dated at about 25,000 years old, to the most recent creations, thought to be about 11,000 years old. One of the principal figures in this analysis was Henri Breuil, who believed Cro-Magnon art represented a form of magic intended to bring good luck to hunters or, perhaps more importantly, protect them from being injured by their prey. Breuil's theory was challenged by Andre Leroi-Gourhan, who placed more importance on the position of an individual painting in a cave and its supposed sexual symbolism. Despite their differences of opinion, both men were in complete agreement about one thing: that early cave art was primitive, and that later creations showed ever-increasing sophistication and flair.

Then Chauvet was discovered, and its masterworks hailed as some of the finest examples of parietal art ever found. This was a pinnacle of Cro-Magnon creativity, it was agreed. "The esthetic and artistic quality of the art is exceptional," said Jean Clottes.[5] But how old was it? Geochronologists scraped flecks of pigment from the cavern walls and took the samples to their laboratory, where the paintings were dated – at 31,000–33,000 years old. "It's not possible," Jean Clottes told them. But the geochronologists stuck to their guns and the dates were verified by two other laboratories.[6] Jean-Marie Chauvet, Eliette Brunel-Deschamps, and Christian Hillaire had not just found one of the best examples of parietal art. They had discovered the world's oldest known art gallery – a collection of paintings whose magnificence and antiquity overturned established ideas about the evolution of human creativity.

A wall of panthers and bisons from the Grotte Combe d'Arc at Chauvet. Such works have been hailed for their "astounding" sensitivity and perception by art critics.

"As the oldest known cave, but with art equal in quality and sophistication to anything that came later, Chauvet gives the lie to previous beliefs in progressive chronology," says Gould.[7] And the art critic John Berger would agree: "What makes the age of these works astounding is the sensitivity of perception they reveal. The thrust of an animal's neck or the set of its mouth or the energy of its haunches were observed and re-created with a nervousness and control comparable to what we find in the works of a Velázquez or a Brancusi. Apparently art did not begin clumsily. The eyes and hands of the first painters and engravers were as fine as any that came later. There was a grace from the start."[8]

"There was a grace from the start." This sentence says it all, although its elegant simplicity hides some challenging issues – because the sudden appearance of works like those found at Chauvet causes considerable anthropological headaches. We are fairly sure Cro-Magnons were responsible for these paintings. Neanderthals were by this time dying out in their last European lairs and, in any case, had shown no previous inclination to contrive art of such a high quality. The trouble is, neither had *Homo sapiens* – either as Cro-Magnons in Europe or as their predecessors in Africa. The 90,000-year-old tools found by Yellen and Brooks at Katanda in the Congo (see Chapter Seven) were superbly crafted, but they cannot compare with the symbolic sophistication, intensity, or grace of the Chauvet cave art.

Indeed, it has become clear that there was very little creative difference between modern humans and Neanderthals until the very end of their time together on this planet. Their cultural similarity is demonstrated clearly by remains found in the caves of Skhul and Qafzeh in Israel. Here, archaeologists have found a plethora of Mousterian implements along with the bones of both *Homo sapiens* and *Homo neanderthalensis*. All are about 100,000 years old. The Levant was probably where these two species first met. The former had begun to spread out of Africa; the latter had moved east as the climate grew colder in Europe. And as their populations ebbed and flowed with the changing climate, they passed through the same terrain and lived in the same caves, although not necessarily at exactly the same time. For long periods, both species were clearly using the same tools, apparently in the

TOP: A reconstruction of a Cro-Magnon artist at work on a cave painting.

ABOVE: A depiction of two rhinoceroses from the Grotte Combe d'Arc cavern at Chauvet.

same way. They occupied roughly the same ecological niche, hunting antelope and gathering fruit and vegetables. Both left evidence of their presence through the interment of their dead, but neither displayed outstanding artistic tendencies. Both seemed to be marking time – happily or not, we do not know.

Then, about 40,000 years ago, a cultural explosion swept through *Homo sapiens*. Tool kits leapt in sophistication in a form called Aurignacian implements that included bone spearheads, fishhooks, and harpoons. Works of art appeared, including sculptures, paintings, and musical instruments. Houses were built – one found in the Ukraine was made entirely of mammoth bone – and minerals, stones, and beads were exchanged. Ivory, antlers, marine shells, limestone, jet, hematite, and other stones were used to make ornaments, some of them selected from quarries that were situated hundreds of miles from their point of manufacture. And, of course, artists began painting stunning animal murals in caves in southern Europe. As John Pfeiffer puts it: "Art came with a bang as far as the archaeological record is concerned."[9]

ABOVE : A harpoon carved from an antler and dated at between 11,000 and 18,000 years old.
ABOVE RIGHT: The Lady of Brassempouy, a 25,000-year-old sculpture carved from mammoth ivory.

And this artistic big bang was certainly extraordinary. Signs of mankind's first intellectual flowering date back to 2.5 million years ago, when hominids – possibly *Homo habilis* – left stone tools strewn across the Olduvai Gorge. And yet, in all the intervening years, no significant improvement had been made, other than "a new type of stone tool here, the introduction of hearths or rudimentary shelters there," as Ian Tattersall points out. Then, abruptly, *Homo sapiens* went on an intellectual rampage. "Counting as one day the time since the first stone tool was made, it is only in the last 20 minutes that we begin to pick up archaeological evidence of the unique modern human sensibility, with its creativity, symbolism, and spirit of constant enquiry and innovation. But when we do, it is with a vengeance," says Tattersall.[10]

Mankind was suddenly on the make, and the Neanderthals simply wilted before the onslaught. But why did we invent culture and not the Neanderthals, and why did we leave it so late? These questions are extraordinarily difficult to answer, despite their clear importance to our story. The problem is that this cultural explosion was not apparent at the beginning of our ancestors' African exodus and did not manifest itself until *Homo sapiens* was well into its great diaspora. Forty thousand years ago, our species had reached Australia and was moving west into Europe. It had spread across the world, and only then decided to erupt culturally. We see the beginnings of rock art in Australia and, not long afterward, the paintings of Chauvet. Only some type of slow-acting neural transformation in the human brain seems able to explain this sudden, disparate flowering of imagination and intellect, resulting in a creative revolution that seems to have emerged in separate, innovative outpourings across Africa, Asia, and Europe.

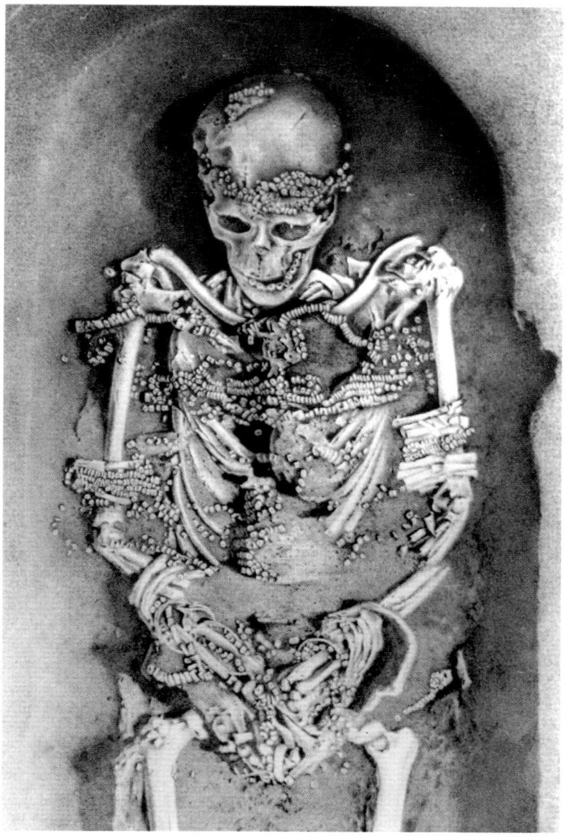

The evidence for this cultural cascade is undeniable. It includes masterpieces such as the Lady of Brassempouy, an exquisitely crafted sculpture of a woman's head carved from mammoth ivory and dated at 25,000 years old. Equally ancient are the Venus figurines, female statuettes with strange, exaggerated breasts and buttocks, that have been found at several sites in Europe. But perhaps most impressive are the 28,000-year-old remains of two children and an adult that archaeologists uncovered at Sungar in Russia. Their bodies were decorated with thousands of ivory beads – so many that it would have taken more than 10,000 hours to make these necklaces and pendants, archaeologists have calculated. Only persons of immense importance would have been considered worth such funereal acclaim. Perhaps the adult was a great warrior. But what could the children have done in their short lives to deserve such honor? The answer is: very little. The only reason their tiny corpses could have been so lavishly adorned was because status was now being inherited rather than achieved. The stratification of society and the division of labor had clearly begun among *Homo sapiens*.

One of the Sungar skeletons, still covered in draperies of ivory beads which had been sewn into its clothes.

This takes us to an important issue: the function of *Homo sapiens'* new-found creativity. We did not invent art for art's sake – it was an innovation that must have had a definite evolutionary purpose. These were hard times, particularly in Ice Age Europe: not the paleolithic equivalent of the Left Bank. Art emerged only because it

HOUSE OF THE SPIRITS

Cave paintings have been found throughout the world: in Brazil, in Australia, in Europe, and in Africa. Some are thousands of years old, but others are very recent, and they give scientists an important clue to the purpose of this form of art. Depictions of animals, such as elands, on the rock walls of Game Pass in South Africa carried out relatively recently have been found to bear striking similarities with the "healing dances" of various African tribesmen. During the healing dance, participants go into a form of hypnotic trance. When asked to recall their experiences afterward, tribal members described images like the ones that had been painted on the walls of Game Pass. In other words, cave art may reflect the trances or rituals in which the spirit of an animal – such as the eland – is raised in attempts to cure the sick.

This idea is stressed by archaeologist Professor David Lewis-Williams of Witwatersrand University, Johannesburg. He points out that constellations of dots and lines accompany many cave paintings – both ancient French and Spanish ones, and also more recent South African ones. These geometric patterns

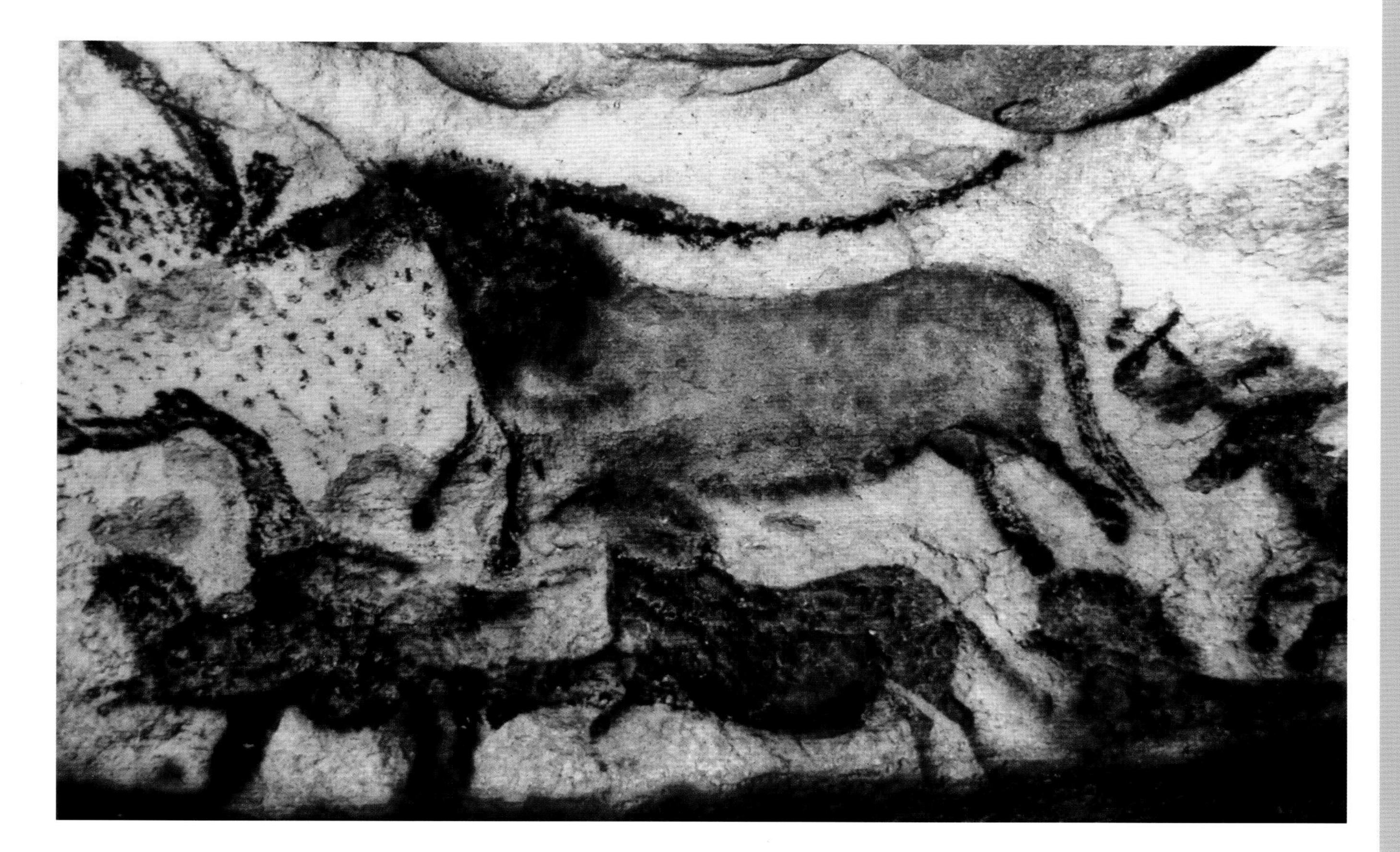

are also reported by individuals who undergo trances. In other words, the great works of art of the Cro-Magnons in places such as Chauvet and Lascaux may be the images of hypnotic reveries, dreams that were considered to be of profound importance to the well-being of the tribe. African bushmen believe the rock walls of caves are a kind of membrane between themselves and the spirit world, says Lewis-Williams. And so possibly did the Cro-Magnons. As he says: "These people were attempting to enter a spirit world."

FAR LEFT: Modern parietal art from Game Pass in South Africa. MIDDLE: A silhouette of a hand that was probably made using a prehistoric "spray gun." Paint was sucked into a thin tube (such as an animal bone) and then blown out as a fine aerosol of saliva and pigment over the hand of the artist or a novice involved in a ceremony within the cave at Peche-Merle, France. ABOVE: The walls of Lascaux in the Perigord, perhaps the most famous of all caverns decorated by the Cro-Magnons. Discovered in the 1940s, the cave boasts hundreds of magnificent murals and became a popular tourist attraction in the 1950s. The huge number of visitors triggered damaging changes to the temperature and humidity, which caused algae to grow on the paintings, so it was decided to close the cave to the public. An underground replica, duplicating its vast halls of art, has recently been opened.

had a role to play in our society, and in the case of the Sungar corpses, we have a hint of what that role might have been. Those decorations gave a symbolic underpinning to society. They defined the status of the individuals that they enveloped, and thus acted as a sort of glue that held together the various strata of a community. For their part, the Venus figurines probably had a key religious function. But what about caves like Chauvet? Apart from their artistic magnificence, what possible social purpose could they have had?

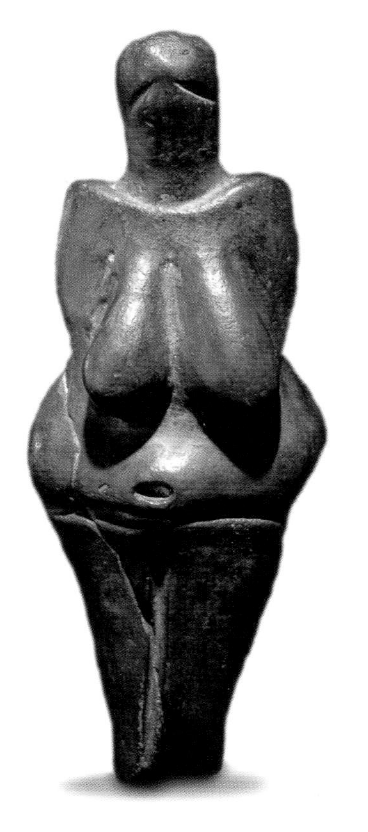

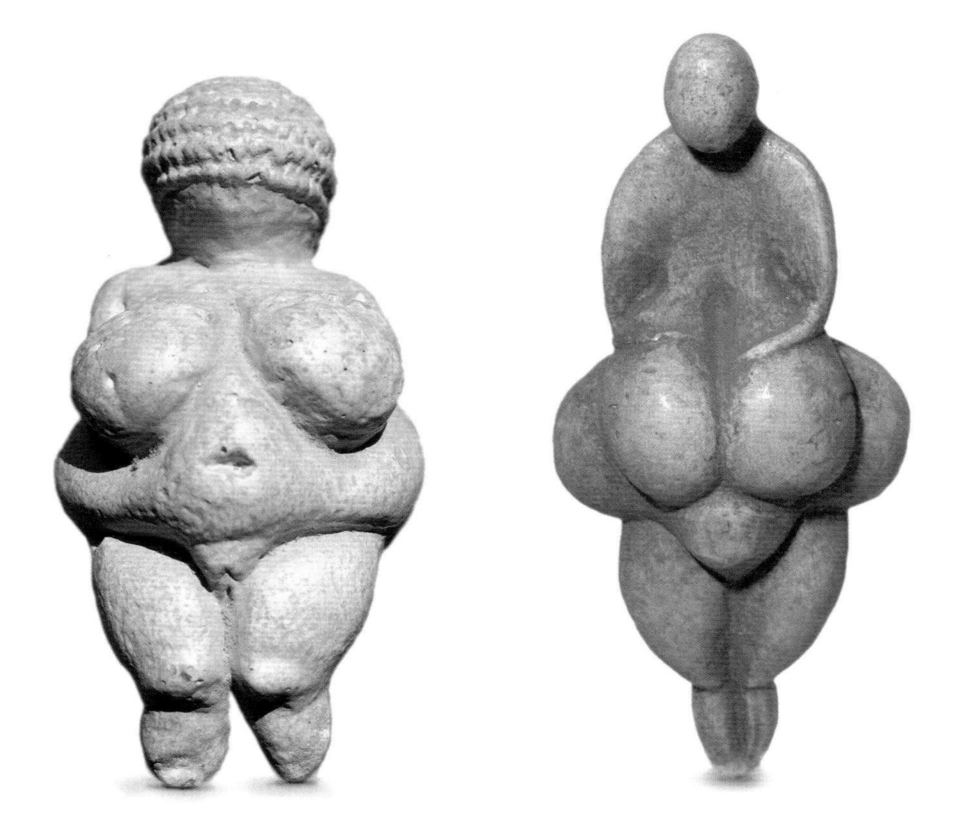

Venus figurines. Statues such as these have been found at several sites in Europe. The earliest are thought to be more than 30,000 years old and generally believed to be some form of fertility symbol.

Again, it is a difficult question to answer, although there are clues, as Nigel Hawkes has outlined in *The Times*. "Some [caves] show many footprints, usually of young people, and almost invariably barefoot. Given the temperatures outside, they must have had shoes or moccasins, but left them at the cave entrance." This evidence – of the specific involvement of so many young people and of their unshod deportment – has led anthropologists to suggest that the deeper parts of the caves were used for ceremonies of initiation. As Hawkes adds, "In some, the most powerful images were created at the greatest depths, as if they were sanctuaries to be visited seldom."[11] One can envisage the scene: a group of youths being led through underground chambers by their elders as they clutch guttering oil lamps and brands, until they reach an inner, sacred sanctum lined with friezes of lions, panthers, and rhinos. There is chanting, perhaps drumming, and everywhere thick, black smoke and flickering lights. It would have been a terrifying, riveting experience that would have been etched forever in those young minds, and that would have helped to cement the edifice of tribal life.

Certainly, it is clear that social interaction had, by now, become profoundly important to human beings. It was a growing web of alliances, more than anything else, that enabled *Homo sapiens* to dominate the world at the expense of the Neanderthals and other hominids. We can see solid evidence of these spreading social tentacles in their trade patterns. While Neanderthal tools are hardly ever found more than 30 miles (50km) from their source, those used by *Homo sapiens* are found up to 200 miles (320km) away, suggesting we had set up sophisticated routes for exchanging goods and for establishing complex familial links. This point is stressed by Clive Gamble. "When you look at the way Neanderthals shared things – like stone tools – you see very little geographic variation, they got all their materials locally. By comparison, modern humans shared objects with other tribes, even though they were often far apart, sometimes hundreds of miles. We networked well, and when times got hard we had kith and kin to run to. The Neanderthals did not. It is like remembering aunties and cousins. We send Christmas cards, but the Neanderthals did not. That doomed them."[12]

Art was critically important in strengthening these social bonds, as was language. Both use symbols to represent ideas and objects, and the latter must also have had a key purpose in helping cement the elaborate relationships that were being formed. Every race and tribe on our planet speaks a language that can be used to communicate extremely complex ideas. A person may utter 40,000 words in a single day, though most will be about "trivial" issues, according to Professor Robin Dunbar of Liverpool University. His assessment of common-room chat at universities found that 86 percent of conversation was about personal relationships and experiences – love lives, TV programs, and jokes.[13] Language is therefore another important social adhesive that would have helped hold tribal life together. It would also have helped

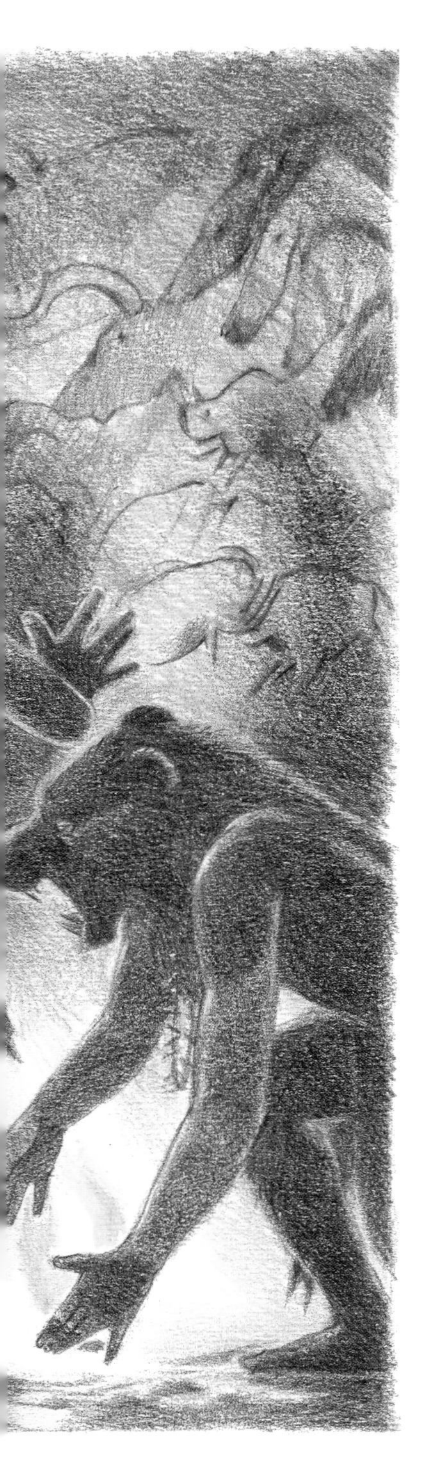

people share survival skills, for complex speech means that "anyone can benefit from the strokes of genius, lucky accidents, and trial-and-error wisdom accumulated by anyone else, present or past," as Harvard University linguist Steve Pinker puts it.[14] By contrast, evidence suggests that Neanderthals communicated less successfully than Cro-Magnons, although it is impossible to be sure. Analysis of the bases of Neanderthal skulls by scientists such as Jeffrey Laitma shows that they were flatter than those of *Homo sapiens*, which suggests their voice boxes were higher and thus not capable of producing the full range of sounds that we can. Conclusive evidence has yet to be found.

And then there is the vexed topic of consciousness, an issue that continues to perplex philosophers and psychologists. Was *Homo sapiens* uniquely gifted in its ability to scrutinize its own mental operations? It is impossible to answer such a question, although the psychologist Nicholas Humphrey believes that consciousness, as exhibited by *Homo sapiens*, may have played a critical role in helping us assemble our labyrinthine relationships. The "depth, complexity, and biological importance" of our social bonds, which "far exceed those of any other animal," would be impossible without our capacity for self-reflexive thought, he says.[15]

In short, it seems that *Homo sapiens* – with better linguistic power, symbolic intellect, and consciousness – slowly took an increasingly firm grip of the terrain, passing on and receiving information about the best hunting land and forests where fruits and berries could be gathered, exchanging and trading in tools, and generally keeping in touch.

OPPOSITE: An artist's impression of an initiation ceremony in a cave painted with scenes such as those at Chauvet. It is believed that such caves were sacred places, used for ritual ceremonies and rites.

Neanderthals, by contrast, continued to adopt a more individualistic response to the rigors of paleolithic life. They hunted harder in smaller groups. "Neanderthals were hominids in the wrong place at the wrong time," says Leslie Aiello. "*Homo sapiens* evolved so that its members expended less energy. We are the economy model."[16] Cheap to run and easy on maintenance, *Homo sapiens* flourished. Isolated and starved of resources, the Neanderthals died out "with a whimper, rather than a bang," as Chris Stringer puts it. This point is backed by Jean-Jacques Hublin. "Neanderthals liked to move around, but returned to favorite caves when times got hard. Slowly groups would find that when they went back to those caves they had been taken over by spreading tribes of *Homo sapiens*. They ran out of places to hide."

Not every paleontologist agrees that *Homo sapiens* was intellectually or culturally superior to the Neanderthals. Some, such as Joao Zilhao, discoverer of the Neanderthal-human half-breed that we encountered in the last chapter, argue that

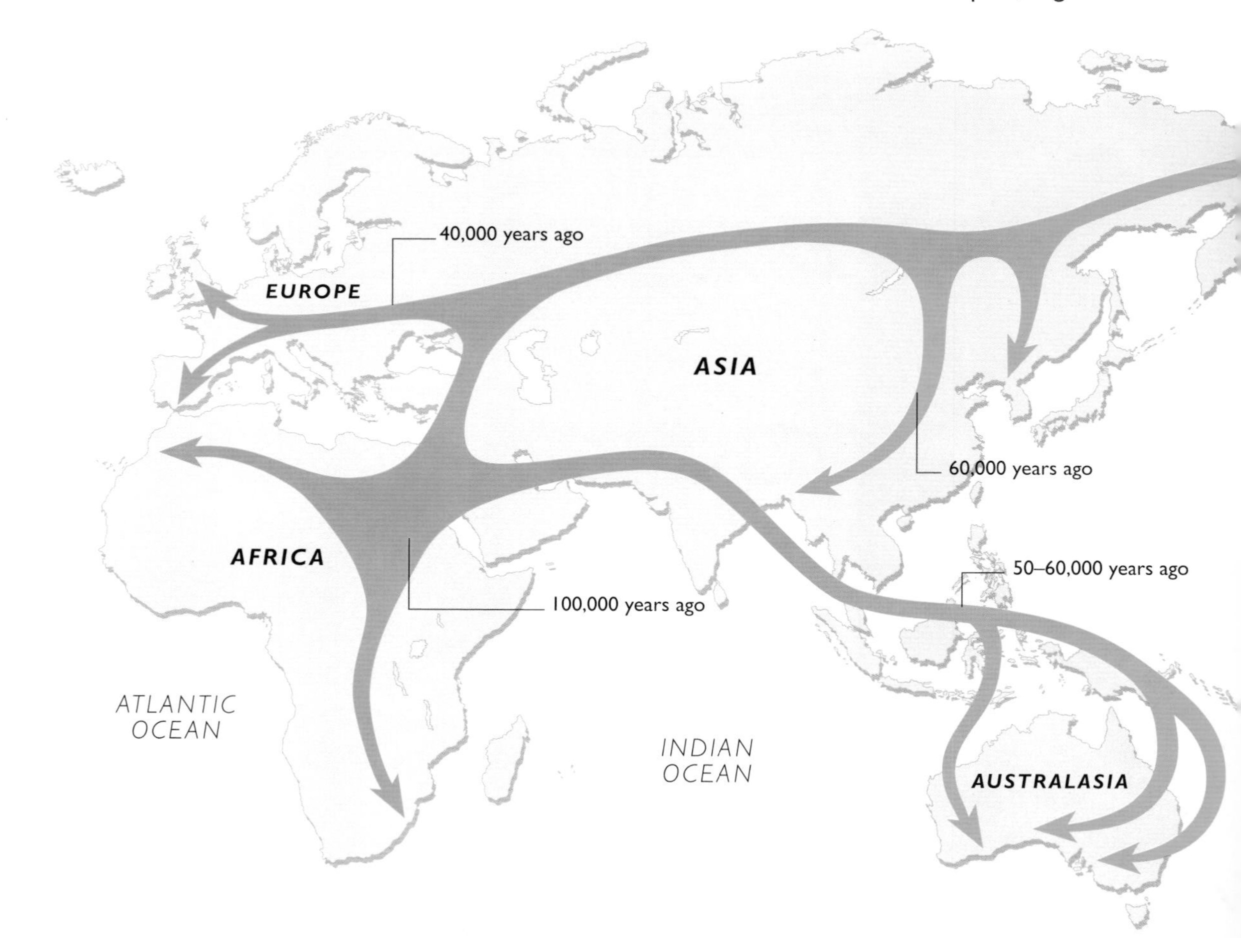

modern humans triumphed for more basic reasons. "There is an idea that modern humans emerged out of Africa like the Chosen People," he says. "Their arrival is portrayed almost as a biblical event, that the golden ones replaced debased Neanderthals. This is nonsense." Zilhao says most evidence for the supposed superiority of *Homo sapiens* – cave paintings, bone tools, sophisticated necklaces, and ornaments – has been poorly dated. The emergence of these artifacts did not coincide with modern humans' arrival in Europe, he argues. Neanderthals were making them before *Homo sapiens* appeared on the scene. "There was virtually no intellectual difference between us and them. We probably bred more and simply outnumbered them."[17] Most paleontologists disagree, however. They acknowledge that, at the very end of their time in Europe, Neanderthals were making quite advanced tools, but they argue that this was probably a result of Neanderthals copying innovations brought in by modern humans. "We did not take over Europe

This map shows the emergence of *Homo sapiens* from Africa and its routes and estimated times of arrival around the world as it gradually dominated the planet.

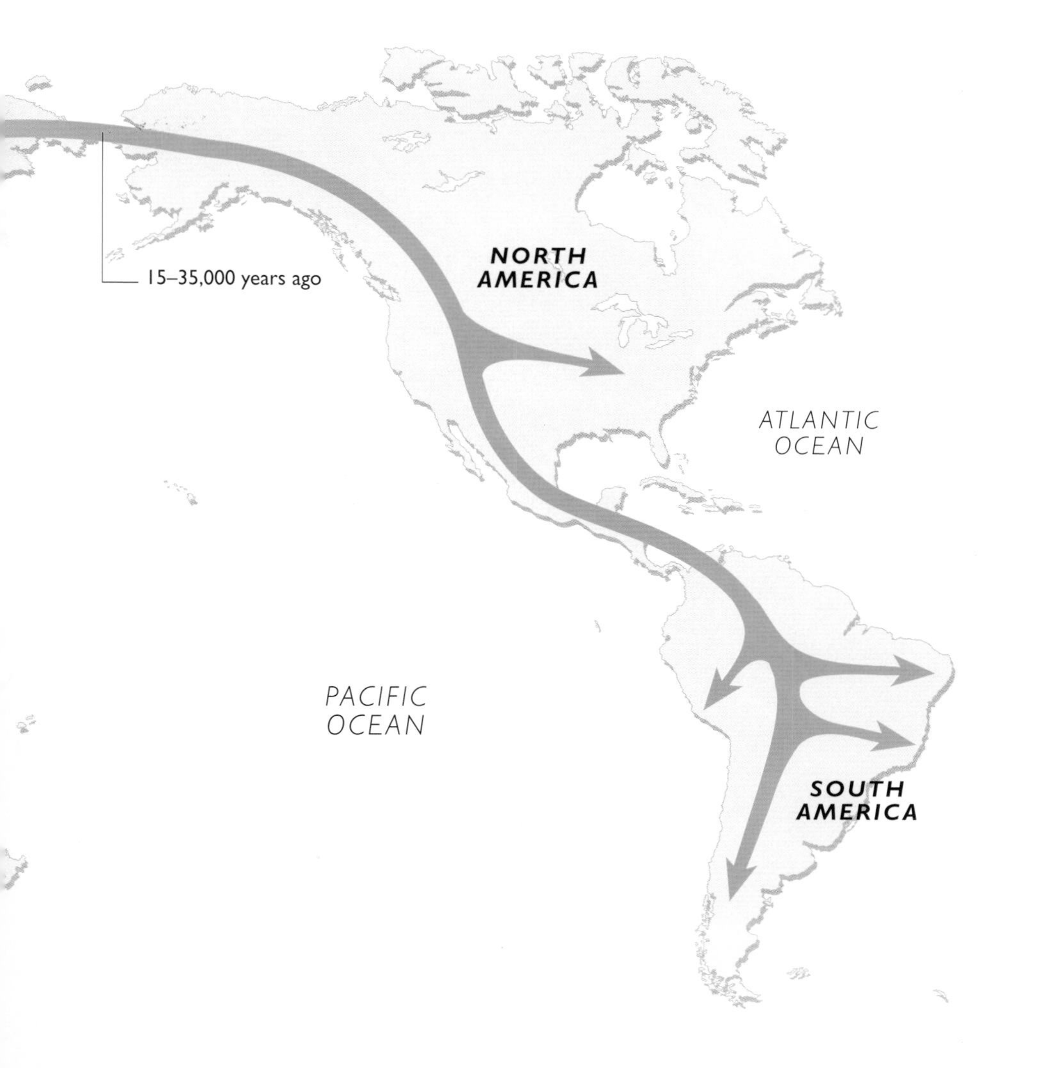

FOOTPRINTS IN THE SANDS OF TIME

Mankind's migration around the world has been traced in some detail by scientists, starting with our exodus from Africa almost 2 million years ago and stretching into more recent times, when modern humans reached the remotest corners of the globe. Many of the details of this diaspora have been unraveled, but one region has provided paleontologists with an unprecedented number of puzzles: Southeast Asia and Australia.

Consider some of the findings of scientist Carl Swisher, whose pioneering research did so much to push back the date of the emergence of *Homo erectus* out of Africa and into Europe and Asia to almost 2 million years before the present. He has examined *Homo erectus* fossils – at Ngandong in Indonesia – that were previously thought to be between 100,000 and 300,000 years old. And this time he has redated them as being far younger: about 50,000 years old. Yet it was around this time that modern humans were making their entry into this part of the world. Swisher's work therefore suggests, incredibly, that *Homo sapiens* and *Homo erectus* may have coexisted for a time in the area.

And then there is the question of mankind's arrival in Australia. Most researchers had assumed that *Homo sapiens* first appeared on the continent about 60,000

years ago. Although ice ages created land bridges between islands, there has never been a link between Asia and Australia in the last 2 million years. Since scientists always assumed that *Homo erectus* never mastered seamanship, Australia was unoccupied until modern humans crossed from Asia in boats. (*Homo sapiens* seem to have been the world's first sailors.)

It has, however, been claimed that some western Australian rock art is 75,000 years old – giving *Homo sapiens* a tight schedule for reaching the continent after emerging from Africa 25,000 years earlier. One group of researchers even claimed to have found evidence of human occupation of Australia 116,000 and 176,000 years ago. But rock art is notoriously difficult to date and these ages have been seriously questioned. However, red ochre, a pigment used in ancient painting, has been found in the oldest well-dated Australian sites, suggesting that at least some of the rock art may indeed be 50,000–60,000 years old.

Australian rock art has remarkable similarities to ancient cave art such as that at Chauvet and Lascaux in France. The image below shows rock art from the Northern Territory that is at least 40,000 years old – around 5,000 years older than Chauvet – but other examples have been found in Australia that are 75,000 years old.

by luck," says Paul Mellars, "We were smarter."[18] Nor is there anything wrong with current dating of European fossils for this period, he adds. Ian Tattersall is similarly convinced that *Homo sapiens* possessed an extra string to its intellectual bow: "Cro-Magnon behavior represented a quantum break with anything previously seen in Europe."[19]

Whatever the cause, the end result was world domination. By 30,000 years ago, give or take the odd millennium, *Homo sapiens* had become the only species of hominid left on Earth – the first time such a monopoly had occurred for millions of years. It remains one of the more philosophically awkward aspects of our prehistory, as Carl Swisher emphasizes. "A lot of people have a problem in dealing with the idea that our total domination of the planet is so recent, and that for millions of years of our prehistory there were so many other different types of human beings. It is a difficulty that is largely religious in nature. Even atheists or agnostics have become inculcated with the idea that there is only one God, and that we – *Homo sapiens* – are made in his image, not a lot of other hominids, like the Neanderthals, as well. Nevertheless, we have to face the fact that we have only been home alone for a very brief period."[20]

That domination was achieved by tribes of hunter-gatherers who displayed a rich creativity and strong social bonding – the blessings and curses of humanity. Together, these attributes have taken us to the Moon, a *tour de force* of intellectual effort and rigorous teamwork, but they have also allowed us to wage global wars, deploying great inventiveness and intense social cohesion, albeit aimed at fellow members of our species. This is the legacy of the Cro-Magnon. They are us, as Gould has said, and our fascination with their handiwork is entirely understandable. When we look at them, we see ourselves – newly minted from Africa, but shorn of several millennia's accumulated technological prowess.

One mystery abides, however. We still have very little idea of the exact location in which *Homo sapiens* first appeared around a quarter of a million years ago. More importantly, we do not know what forces shaped our ancestors and gave them the final piece of evolutionary spin-doctoring that allowed them to take over the world. Somewhere, *Homo sapiens* evolved features that would, by good luck, ensure its domination of our planet. What they were, and why they evolved in that particular part of the world, we still have no idea. All we do know is that in one region of sub-Saharan Africa, a special set of circumstances prevailed, and that the successors of the hominids who were subject to such rigors were the people who inherited the Earth.

What followed the Cro-Magnons' arrival is a little easier to establish, however. Having slowly taken over the Old World and Australasia, mankind spread across Beringia – the bridge of land that once straddled the straits between modern Russia and Alaska – and passed down into North and then South America. In the process, dozens of ancient species were eradicated: saber-toothed cats, mammoths, mastodons, four-horned antelopes, llamalike mammals, and many others. By about 11,000 years ago, all the main landmasses of the world had been conquered (apart from Antarctica). Eventually even the smallest, remotest islands of the Pacific were settled.

Around the same time that America was conquered, mankind began an experiment that was to have the most profound implications: agriculture. Foragers would have noticed when they returned to a favorite site that discarded seeds and plants had begun to grow. A few individuals began to intervene, first to create reliable reserve larders in case wild food supplies failed, then to provide the principal source of nourishment. Within a short time, wheat was being cultivated in the Middle East, rice in China, corn in South America, and sorghum, millet, and yams in western Africa. The whole business was further aided by significant climatic changes, which had ended the last ice age and brought hotter, moister weather to the world. Agriculture allowed more people to thrive on a given acreage, although these individuals were probably less well nourished than their hunter-gatherer predecessors, who would have had a richer choice of foods to eat. Nevertheless, those who followed the old ways found themselves outnumbered and in territorial retreat – a process that continues to this day. "Within the next decades, the few remaining bands of hunter-gatherers will abandon their ways, disintegrate, or die out, thereby ending our millions of years of commitment to the hunter-gatherer lifestyle," warns anthropologist Jared Diamond.[21]

The idea of farming spread, and along with it the notion of land ownership. The excess manpower supported by wheat or rice was used to form armies that fought over resources, and writing was invented to help keep agricultural records. The latter development allowed humans to store and accumulate knowledge, triggering the eventual evolution of technology. Not every continent was equally blessed, however. Eurasia – formed by Europe and the Middle East – had the richest agricultural resources. These included cattle, sheep, goats, poultry, horses, wheat, and barley. "The

The churn (right) and the grinding stone (below), dating from around 4500–5000 BC, are examples of early agricultural tools found in Jerusalem.

unequal distribution of wild ancestral species among the continents became an important reason why Eurasians, rather than people of other continents, were the ones to end up with guns, germs, and steel," says Diamond.[22] Rich in agriculture, the people who were to make up the Western nations simply had too powerful a head start in technology for others to compete with.

This point is a crucial one, for as we have seen, the Out of Africa theory makes it clear that *Homo sapiens* is a startlingly homogenous species. Racial differences are more apparent than real, the main one – skin color – being merely an evolutionary response to ultraviolet radiation in sunlight. So why, then, is there such a disparity of resources between races today? Why was it that the Spaniards conquered the Incas and not the other way around? And why do Western powers own Cruise missiles and laser-guided bombs, but not the native peoples of other countries? Past explanations – that Westerners were instinctively more "vigorous" or "ambitious" – reek of racism, particularly in the light of our new knowledge of mankind's recent African origin. By contrast, Diamond's answer – that the players were equal but the playing fields tilted – fits far more accurately with what we now know about human evolution.

And thus the inheritance of the Cro-Magnons has been distributed. However, it is not the purpose of this book to explain in exact detail how, or why, it has been left in such an unfair state. That is a matter for historians and politicians. The paleontologist has merely to define in broad terms how we arrived at our present state. The devil, as always, is in the detail –

a concern for chroniclers of our recent past, not for those interested in unraveling our evolution from apeman ancestors.

That latter story has been this book's prime focus and is now complete. We have learned how the first hominids split from the common ancestors that we share with chimpanzees, and we have followed their first hesitant steps onto the savanna and the cinder beds of Laetoli. This was followed by the demise of the Nariokotome boy, the mysterious deaths of the people of La Sima, and the lonely demise of the last Neanderthals, until we eventually arrive at *Homo sapiens*. It is a journey of three key phases. The first – that of *Australopithecus* – was carried out by creatures defined by their bipedalism. Then there was *Homo erectus*, marked by its tool-making and omnivorous diet – habits that led the species to cross the Old World. And, finally, there was the arrival of modern humans, whose intellectual prowess and social cohesiveness have left them as Earth's only hominids. The common theme of this evolutionary epic is that, in the face of increasing climatic uncertainty, human beings evolved more and more flexible responses – widening their territorial ranges through upright locomotion, expanding their diets, opening up new environmental niches through tool manufacture, and increasing and strengthening their social groupings to maximize mutual support.

The end result is a species that is now capable of searching for an explanation for its own appearance on Earth. As the British biologist Richard Dawkins states, "Intelligent life on a planet comes of age when it first works out the reason for its own existence."[23] By that definition, we have arrived.

NOTES

CHAPTER ONE

1. M. Leakey (1979). Footprints in the Ashes of Time. *National Geographic*, April.
2. Conversation with the author (1999).
3. *Ibid.*
4. M. Leakey (1979). Footprints in the Ashes of Time. *National Geographic*, April.
5. N. Agnew & M. Demas (1998). Preserving the Laetoli Footprints. *Scientific American*, September, 26–37.
6. M. Leakey (1979). Footprints in the Ashes of Time. *National Geographic*, April.
7. Conversation with the author (1999).
8. M. Leakey (1979). Footprints in the Ashes of Time. *National Geographic*, April.
9. N. Agnew & M. Demas (1998). Preserving the Laetoli Footprints. *Scientific American*, September, 26–37.
10. I. Tattersall (1993). *The Human Odyssey*. Prentice Hall, New York.
11. *Ibid.*
12. I. Tattersall (1998). The Laetoli Diorama. *Scientific American*, September, 35.
13. M. Leakey (1979). Footprints in the Ashes of Time. *National Geographic*, April.
14. *Ibid.*
15. N. Agnew & M. Demas (1998). Preserving the Laetoli Footprints. *Scientific American*, September, 26–37.
16. Conversation with the author (1999).
17. Leakey buried the steps using local river sand when she closed down her site investigation in 1979. The aim was to protect the Laetoli footprints for posterity. However, in doing so, acacia seeds were inadvertently introduced to the trackway. By 1992, scientists had become so concerned about the damage that roots of the acacia might be inflicting on this precious piece of land that the sand was dug up, the steps exposed, and a new burial mound built over the footprints. See N. Agnew and M. Demas (1998). Preserving the Laetoli Footprints. *Scientific American*, September, 26–37.
18. M. Leakey (1979). Footprints in the Ashes of Time. *National Geographic*, April.
19. R. Leakey (1994). *The Origin of Humankind*. Weidenfeld & Nicolson, London.
20. M. Leakey (1979). Footprints in the Ashes of Time. *National Geographic*, April.
21. R. Lewin (1999). *Human Evolution: an Illustrated Introduction*. Blackwell.
22. S. J. Gould (1980). Our Greatest Evolutionary Step. *Panda's Thumb*. Penguin, London.
23. O. Lovejoy quoted in R. Leakey (1994). *The Origin of Humankind*. Weidenfeld & Nicolson, London.
24. D. Johanson & B. Edgar (1996). *From Lucy to Language*. Weidenfeld & Nicolson, London.
25. R. McKie (1993). Two Legs Best for Brainy Ancestor. *Observer*, November 14.
26. Conversation with the author (1999).
27. Meave Leakey (1995). Exploration in East Africa Reveals Apelike Creatures that Walked Upright Four Million Years Ago. *National Geographic*, September, 38–51.
28. *Ibid.*
29. I. Tattersall (1998). *Becoming Human*. Oxford University Press.
30. Conversation with the author (1999).
31. Conversation with the author (1999).
32. D. Johanson (1996). Face to Face with Lucy's Family. *National Geographic*, March, 96–117.
33. D. Johanson (1976). Ethiopia Yields First Family of Early Man. *National Geographic*, December, 791–811.
34. *Ibid.*
35. *Ibid.*
36. *Ibid.*
37. D. Johanson (1996). Face to Face with Lucy's Family. *National Geographic*, March, 96–117.
38. This fairly neat picture has recently been clouded by the discovery of a new australopithecine – in Chad, far to the west of all other ancient hominid sites. The discovery, made by Michel Brunet of the University of Poitiers, has been named *Australopithecus bahrelghazali*, and its presence, hundreds of miles west of the Rift Valley, is difficult to reconcile with the East Side Story theory of hominid evolution. However, further studies still have to be completed.
39. D. Johanson (1996). Face to Face with Lucy's Family. *National Geographic*, March, 96–117.
40. Quoted by R. Lewin (1995). Bones of Contention: Two Recent Finds of Early Human Fossils have Triggered a Revolution in the Way Anthropologists Think about Evolution. *New Scientist*, November 4.

CHAPTER TWO

1. C. Tudge (1995). Human Origins – A Family Feud: Have We Discovered All the Ancestors of *Homo sapiens*? *New Scientist*, May 20.
2. D. Johanson (1996). Face to Face with Lucy's Family. *National Geographic*, March, 96–117.
3. Quoted in R. Lewin (1999). *Human Evolution, an Illustrated Introduction*. Blackwell.
4. *Ibid.*
5. *Ibid.*
6. Conversation with the author (1994).
7. I. Tattersall (1998). *Becoming Human*. Oxford University Press.
8. S. J. Gould (1998). Our Unusual Unity. *Leonardo's Mountain of Clams and the Diet of Worms*. Jonathan Cape, London.
9. R. Dart. *Adventures with the Missing Link*. Quoted in J. Reader (1981). *Missing Links: The Hunt for Earliest Man*. Collins, London.
10. The name *Australopithecus africanus* was controversial, not least because it was a mix of Greek and Latin. *Australo* is Latin for "southern" and *pithecus* is latinized Greek for "ape." Such a combination engendered antagonism. "It is generally felt that the name *Australopithecus* is an unpleasing hybrid as well as being etymologically incorrect," *Nature* announced on March 28, 1925, while the British anthropologist Sir Arthur Smith Woodward described the mixture as "barbarous." See J. Reader (1981). *Missing Links: The Hunt for Earliest Man*. Collins, London.
11. C. Stringer and R. McKie (1996). *African Exodus*. Jonathan Cape, London.
12. Quoted in J. Reader (1981). *Missing Links: The Hunt for Earliest Man*. Collins, London.
13. C. Darwin (1881). *The Descent of Man*. John Murray, London.
14. J. Reader (1981). *Missing Links: The Hunt for Earliest Man*. Collins, London.
15. *Ibid.*
16. Quoted in J. Reader (1981). *Missing Links: The Hunt for Earliest Man*. Collins, London.
17. Quoted in D. Johanson & B. Edgar (1996). *From Lucy to Language*. Weidenfeld & Nicolson.
18. J. Reader (1981). *Missing Links: The Hunt for Earliest Man*. Collins, London.
19. This vegetarian specialization has led many experts to classify both *robustus* and *boisei* as a different genus, *Paranthropus* (near man).
20. L. Leakey (1960). Finding the World's Earliest Man. *National Geographic*, June, 420–435.
21. J. Reader (1981). *Missing Links: The Hunt for Earliest Man*. Collins, London.
22. V. Morell (1996). *Ancestral Passions: The Leakey Family and the Quest for Humankind's Beginnings*. Touchstone, New York.
23. L. Leakey (1960). Finding the World's Earliest Man. *National Geographic*, June, 420–435.
24. R. Lewin (1987). *Bones of Contention*. Simon and Schuster, New York.
25. Quoted in V. Morell (1996). *Ancestral Passions: The Leakey Family and the Quest for Humankind's Beginnings*. Touchstone, New York.
26. *Ibid.*
27. B. Wood & M. Collard (1999). The Human Genus. *Science* 284, 65–71.
28. A. Walker & P. Shipman (1996). *Wisdom of the Bones*. Weidenfeld & Nicolson, London.
29. Conversation with the author (1999).
30. Bernard Wood has recently reclassified both *Homo habilis* and *Homo rudolfensis* as *Australopithecus habilis* and *Australopithecus rudolfensis*. B. Wood & M. Collard (1999). The Human Genus. *Science* 284, 65–71.
31. W.H. Calvin (1998). The Emergence of

Intelligence. *Scientific American*. Special issue: Exploring Intelligence. Winter.
32. Conversation with the author (1999).

CHAPTER THREE

1. M. Leakey (1995). The Farthest Horizon. *National Geographic*, September, 38–51.
2. A. Walker & P. Shipman (1996). *The Wisdom of the Bones: In Search of Human Origins*. Weidenfeld & Nicolson, London.
3. R. Leakey & R. Lewin (1992). *Origins Reconsidered*. Little, Brown & Co., London.
4. A. Walker & P. Shipman (1996). *The Wisdom of the Bones: In Search of Human Origins*. Weidenfeld & Nicolson, London.
5. R. Leakey & R. Lewin (1992). *Origins Reconsidered*. Little, Brown & Co., London.
6. A. Walker & P. Shipman (1996). *The Wisdom of the Bones: In Search of Human Origins*. Weidenfeld & Nicolson, London.
7. *Ibid*.
8. R. Leakey & R. Lewin (1992). *Origins Reconsidered*. Little, Brown & Co., London.
9. A. Walker & P. Shipman (1996). *The Wisdom of the Bones: In Search of Human Origins*. Weidenfeld & Nicolson. London.
10. C. Stringer & R. McKie (1996). *African Exodus*. Jonathan Cape, London.
11. A. Walker & P. Shipman (1996). *The Wisdom of the Bones: In Search of Human Origins*. Weidenfeld & Nicolson, London.
12. R. Leakey & A. Walker (1985). *Homo Erectus* Unearthed: A Fossil Skeleton 1,600,000 years old. *National Geographic*. November, 624–9.
13. A. Walker & P. Shipman (1996). *The Wisdom of the Bones: In Search of Human Origins*. Weidenfeld & Nicolson, London.
14. R. Leakey & A. Walker (1985). *Homo Erectus* Unearthed: A Fossil Skeleton 1,600,000 years old. *National Geographic*. November, 624–9.
15. A. Walker & P. Shipman (1996). *The Wisdom of the Bones: In Search of Human Origins*. Weidenfeld & Nicolson, London.
16. *Ibid*.
17. R. Leakey (1981). *The Making of Mankind*. Michael Joseph, London.
18. Conversation with the author (1999).
19. A. Walker & P. Shipman (1996). *The Wisdom of the Bones: In Search of Human Origins*. Weidenfeld & Nicolson, London.
20. *Ibid*.
21. J. Desmond Clark: quoted at the meeting in his honor, "The Longest Record: the Human Career in Africa," Berkeley, April 1986.
22. Conversation with the author (1999).
23. *Ibid*.
24. Conversation with the author (1999).
25. A. Walker & P. Shipman (1996). *The Wisdom of the Bones: In Search of Human Origins*. Weidenfeld & Nicolson, London.
26. Conversation with the author (1999).
27. Conversation with the author (1999).
28. Conversation with the author (1995).
29. C. Stringer & R. McKie. (1996). *African Exodus*. Jonathan Cape, London.
30. *Ibid*.
31. R. Leakey (1981). *The Making of Mankind*. Michael Joseph, London.
32. A. Walker & P. Shipman (1996). *The Wisdom of the Bones: In Search of Human Origins*. Weidenfeld & Nicolson, London.

CHAPTER FOUR

1. J. Reader (1981). *Missing Links*. Collins, London.
2. S. J. Gould (1988). Men of the Thirty-third Division. *Eight Little Piggies*. Jonathan Cape, London.
3. J. Reader (1981). *Missing Links*. Collins, London.
4. S. J. Gould (1988). Men of the Thirty-third Division. *Eight Little Piggies*. Jonathan Cape, London.
5. Conversation with the author (1999). Also R. Gore (1997). Expanding Worlds. *National Geographic*, May.
6. R. Lewin (1999). *Human Evolution: an Illustrated Introduction*. Blackwell.
7. J. Reader (1981). *Missing Links*. Collins, London.
8. M. Lemonick (1994). How Man Began. New evidence shows that early humans left Africa much sooner than once thought. Did *Homo sapiens* evolve in many places at once? *Time*, March.
9. Conversation with the author (1999).
10. A. Walker & P. Shipman (1996). *The Wisdom of the Bones: In Search of Human Origins*. Weidenfeld & Nicolson, London.
11. Conversation with the author (1999).
12. A. Walker & P. Shipman (1996). *The Wisdom of the Bones: In Search of Human Origins*. Weidenfeld & Nicolson, London.
13. Quoted in M. Lemonick (1994). How Man Began. New evidence shows that early humans left Africa much sooner than once thought. *Time*. March.
14. Quoted in R. Gore (1997). Expanding Worlds. *National Geographic*, May, 84–109.
15. R. Gore (1997). Expanding Worlds. *National Geographic*, May, 84–109.
16. K. Weaver (1985). The Search for our Ancestors. *National Geographic*, November, 560–623.
17. R. Lewin (1999). *Human Evolution: an Illustrated Introduction*. Blackwell.
18. J. Kingdon (1993). *Self-made Man and his Undoing*. Simon & Schuster, London.
19. Conversation with the author (1999).
20. E. Pennisi (1999). Did Cooked Tubers Spur the Evolution of Big Brains? *Science*, March 26.
21. *Ibid*.
22. Conversation with the author (1999).
23. *Ibid*.
24. R. McKie (1998). Granny Power is Secret of Human Survival. *The Observer*, September 20.
25. Conversation with the author (1998)

CHAPTER FIVE

1. R. Kunzig (1997). Atapuerca: The Face of the Ancestral Child. *Discover*, December, 88–101.
2. R. McKie (1997). To hell and back in search of a lost race. *The Observer*, June 1.
3. Conversation with the author, Atapuerca (1997).
4. R. Kunzig (1997). Atapuerca: the Face of the Ancestral Child. *Discover*, December, 88–101.
5. Conversation with the author (1999).
6. R. McKie (1996). Boxgrove Man Goes Back Underground. *The Observer*, October 20.
7. Conversation with the author (1999).
8. I. Tattersall (1998). *Becoming Human*. Oxford University Press.
9. Conversation with the author (1999).
10 Conversation with the author (1999).
11. Quoted in R. Gore (1997). The First Europeans. *National Geographic*, July, 96–113.
12. Conversation with the author (1999).
13. *Ibid*.
14. Quoted in R. Gore (1997). The First Europeans. *National Geographic*, July, 96–113.
15. Conversation with the author (1999).
16. I. Tattersall (1998). *Becoming Human*. Oxford University Press.
17. Quoted in M. Pitts & M. Roberts (1997). *Fairweather Eden*. Century, London.
18. R. Potts (1998). Variability Selection in Hominid Evolution. *Evolutionary Anthropology*.
19. *Ibid*.
20. Conversation with the author (1999).
21. Conversation with the author (1999).
22. Conversation with the author (1999).
23. Conversation with the author, Atapuerca (1997).
24. *Ibid*.
25. *Ibid*.
26. *Ibid*.
27. Conversation with the author, London (1999).
28. *Ibid*.
29. Conversation with the author (1999).
30. R. Kunzig (1997). Atapuerca: the Face of the Ancestral Child. *Discover*, December, 88–101.
31. Conversation with the author, London (1999).
32. *Ibid*.
33. *Ibid*.

NOTES (CONTINUED)

CHAPTER SIX

1. H.G. Wells (1921). *The Grisly Folk.* (Reprinted in H.G. Wells (1958), *Selected Short Stories.* Harmondsworth, Penguin.)
2. J. Reader (1981). *Missing Links.* Collins. London.
3. J. Reader, op. cit. R. Lewin, op. cit.
4. C. Stringer & R. McKie (1996). *African Exodus.* Jonathan Cape, London.
5. J. Shreeve (1995). *The Neanderthal Enigma.* William Morrow, New York.
6. *Ibid.*
7. C. Stringer & R. McKie (1996). *African Exodus.* Jonathan Cape, London.
8. Conversation with the author, Amud (1993), and quoted in C. Stringer & R. McKie (1996). *African Exodus.* Jonathan Cape, London.
9. J. Shreeve (1995). *The Neanderthal Enigma.* William Morrow, New York.
10. R. Gore (1996). Neandertals. *National Geographic,* January.
11. R. S. Solecki (1971). *Shanidar: The First Flower People.* Knopf, New York.
12. C. Stringer & R. McKie (1996). *African Exodus.* Jonathan Cape, London.
13. Conversation with the author. Also R. Gore (1996). Neandertals. *National Geographic,* January.
14. Quoted in R. Gore (1996). Neandertals. *National Geographic,* January.
15. *Ibid.*
16. *Ibid.*
17. J. Shea, quoted in interview with D. Lieberman, Harvard (1994). Also C. Stringer & R. McKie (1996). *African Exodus.* Jonathan Cape, London.
18. K. Schick & N. Toth (1993). *Making Silent Stones Speak.* Weidenfeld & Nicolson, London.
19. W. Golding (1961). *The Inheritors.* Faber, London.
20. A. J. E. Cave & W.L. Straus (1957). Pathology and Posture of Neanderthal Man. *Quarterly Review of Biology* 32, 348–363.
21. J. S. Jones (1993). *The Language of the Genes.* HarperCollins, London.
22. J. Shreeve (1995). *The Neanderthal Enigma.* William Morrow, New York.
23. Quoted in R. Gore (1996). Neandertals. *National Geographic,* January.
24. C. Stringer & R. McKie (1996). *African Exodus.* Jonathan Cape, London.
25. J. Shreeve (1995). *The Neanderthal Enigma.* William Morrow, New York.
26. Conversation with the author, Zafarraya (1994).
27. J. S. Jones (1994). A Brave, New, Healthy World? *Natural History,* June.

CHAPTER SEVEN

1. J. Shreeve (1995). *The Neanderthal Enigma.* William Morrow, New York.
2. J Kingdon (1993). *Self-made Man and his Undoing.* Simon & Schuster, London.
3. The honor of giving the name Eve to this genetic mother-figure is generally accorded to Charles Petit, the distinguished science writer of the *San Francisco Chronicle.* Wilson claimed he disliked the title, preferring instead "mother of us all" or "one lucky mother." Few observers see much difference. Source: M. Brown (1990). *The Search for Eve.* HarperCollins, New York.
4. The Search for Adam and Eve. *Newsweek,* 11 January 1988.
5. C. Stringer & R. McKie (1996). *African Exodus.* Jonathan Cape, London.
6. C. Stringer (1996). Out of Eden: A Personal History. Included in C. Stringer & R. McKie. *African Exodus.* Jonathan Cape, London.
7. D.C. Johanson & B. Edgar (1996). *From Lucy to Language.* Weidenfeld & Nicolson.
8. R. Leakey (1983). *One Life.* Michael Joseph, London.
9. J. Shreeve (1992). The Dating Game. *Discover,* September.
10. C. Stringer & R. McKie (1996). *African Exodus.* Jonathan Cape, London.
11. C. L. Brace (1964). The Fate of the "Classic" Neanderthals: A Consideration of Hominid Catastrophism. *Current Anthropology.*
12. Quoted in J. Shreeve (1995). *The Neanderthal Enigma.* William Morrow, New York.
13. C. Stringer (1996). Out of Eden: A Personal History. Included in C. Stringer & R. McKie. *African Exodus.* Jonathan Cape, London.
14. The Search for Adam and Eve. *Newsweek.* 11 January 1988.
15. A. Templeton (1992). Human Origins and the Analysis of Mitochondrial DNA Sequences. *Science* 255, 737. A. Templeton (1993). The "Eve" Hypothesis: A Genetic Critique and Re-analysis. *American Anthropologist* 95, 51–72.
16. M. Ruvolo, interview with author (1994), Harvard. Quoted in C. Stringer & R. McKie. *African Exodus.* Jonathan Cape, London.
17. L. Cavalli-Sforza, P. Menozzi & A. Piazza (1994). *The History and Geography of Human Genes.* Princeton University Press, New Jersey.
18. Conversation with the author, London (1997).
19. R. McKie (1997). Ancient Genes that Told a Story. *The Observer,* July 13.
20. M. Ruvolo *et al.* (1993). Mitochondrial COII Sequences and Modern Human Origins. *Molecular Biology and Evolution.* D. Woodruff et al.. (1999). *Proceedings of National Academy of Sciences.*
21. S. J. Gould (1998). Our Universal Unity. *Leonardo's Mountain of Clams and the Diet of Worms.* Jonathan Cape, London.
22. S.J. Gould (1994). So Near and Yet So Far. *New York Review of Books,* 24–28. In the Mind of the Beholder. *Natural History,* February, 14–23.
23. Quoted in A. Palmer (1999). Did She Become Woman? *Sunday Telegraph,* April 25.
24. *Ibid.*

CHAPTER EIGHT

1. Quoted in S. Lowry (1997). The Oldest Art Gallery. *The Sunday Telegraph,* June 15.
2. S. J. Gould (1998). Up Against the Wall. *Leonardo's Mountain of Clams and the Diet of Worms.* Jonathan Cape, London.
3. Quoted in T. Patel (1996). Stone Age Picassos. *New Scientist,* July 13.
4. S. J. Gould (1998). Up Against the Wall. *Leonardo's Mountain of Clams and the Diet of Worms.* Jonathan Cape, London.
5. Quoted in T. Patel (1995). Ancient Masters Out Painting in Perspective. *New Scientist,* June 17.
6. Quoted in T. Patel (1996). Stone Age Picassos. *New Scientist,* July 13.
7. S. J. Gould (1996). In the Beginning. *Observer,* April 14.
8. J. Berger (1996). Secrets of the Stone. *Guardian,* November16.
9. J. Pfeiffer (1982). *The Creative Explosion.* Harper & Row, New York.
10. I. Tattersall (1993). *The Human Odyssey.* Prentice Hall, New York.
11. N. Hawkes (1995). The Original Man. *The Times,* June 17.
12. R. McKie (1998). Scientists Squabble Over the Fate of the Last Neanderthals. *The Observer,* 30August.
13. R. Dunbar (1994). *The Times,* February 5.
14. S. Pinker (1994). *The Language Instinct.* Allen Lane.
15. Quoted in I. Tattersall (1998). *Becoming Human.* Oxford University Press.
16. Conversation with the author (1999).
17. R. McKie (1998). Scientists Squabble Over the Fate of the Last Neanderthals. *The Observer,* August 30.
18. *Ibid.*
19. I. Tattersall (1993). *The Human Odyssey.* Prentice Hall, New York.
20. Conversation with the author (1999).
21. J. Diamond (1997). *Guns, Germs and Steel.* Jonathan Cape, London.
22. *Ibid.*
23. R. Dawkins (1976). *The Selfish Gene.* Oxford University Press.

BIBLIOGRAPHY

Brown, M. (1990). *The Search for Eve*. HarperCollins, New York.

Darwin, C. (1881). *The Descent of Man*. John Murray, London.

Dawkins, R. (1976). *The Selfish Gene*. Oxford University Press.

Diamond, J. (1997). *Guns, Germs and Steel*. Jonathan Cape, London.

Golding, W. (1961). *The Inheritors*. Faber & Faber, London.

Goodall, J. (1988 rev. edn.). *In the Shadow of Man*. Weidenfeld & Nicolson, London.

Gould, S. J. (1980). Our Greatest Evolutionary Step. *Panda's Thumb*. Penguin, London.

Gould, S. J. (1988). Men of the Thirty-third Division. *Eight Little Piggies*. Jonathan Cape, London.

Gould, S. J. (1998). Our Universal Unity. *Leonardo's Mountain of Clams and the Diet of Worms*. Jonathan Cape, London.

Johanson, D. C. and Edey, M.A. (1990). *Lucy: The Beginnings of Humankind*. Penguin, London.

Johanson, D. C. and Edgar, B. (1996). *From Lucy to Language*. Weidenfeld & Nicolson, London.

Jones, J. S. (1993). *The Language of the Genes*. HarperCollins, London.

Kingdon, J. (1993). *Self-made Man and his Undoing*. Simon & Schuster, London.

Leakey, R. & Lewin, R. (1992). *Origins Reconsidered*. Little, Brown & Co., London.

Leakey, R. (1981). *The Making of Mankind*. Michael Joseph, London.

Leakey, R. (1983). *One Life*. Michael Joseph, London

Leakey, R. (1994). *The Origin of Humankind*. Weidenfeld & Nicolson, London.

Lewin, R. (1987). *Bones of Contention*. Simon and Schuster, New York.

Lewin, R. (1999). *Human Evolution, an Illustrated Introduction*. Blackwell, Oxford.

Morell, V. (1996). *Ancestral Passions: The Leakey Family and the Quest for Humankind's Beginnings*. Touchstone, New York.

Pfeiffer, J. (1982). *The Creative Explosion*. Harper & Row, New York.

Pinker, S. (1994). *The Language Instinct*. Allen Lane, London.

Pitts, M. & Roberts, M. (1997). *Fairweather Eden*. Century, London.

Reader, J. (1981). *Missing Links: The Hunt for Earliest Man*. Collins, London.

Schick, K. & Toth, N. (1993). *Making Silent Stones Speak*. Weidenfeld & Nicolson, London.

Shreeve, J. (1995). *The Neanderthal Enigma*. William Morrow, US.

Solecki, R.S. (1971). *Shanidar: The First Flower People*. Knopf, New York.

Stringer, C. & McKie, R. (1996). *African Exodus*. Jonathan Cape, London.

Tattersall, I. (1993). *The Human Odyssey*. Prentice Hall, New York.

Tattersall, I. (1995). *The Fossil Trail*. Oxford University Press.

Tattersall, I. (1998). *Becoming Human*. Oxford University Press.

Walker, A. & Shipman, P (1996). *The Wisdom of the Bones: In Search of Human Origins*. Weidenfeld & Nicolson, London.

Wells, H.G. (1921). *The Grisly Folk*. (Reprinted in H.G. Wells (1958), *Selected Short Stories*. Harmondsworth, Penguin.)

PICTURE CREDITS

BBC Books would like to thank the following for providing photographs and for permission to reproduce copyright material. While every effort has been made to trace and acknowledge all copyright holders, we would like to apologise for any errors or omissions.

Key: a above, b below, r right, l left, c centre

Agence Photographique de la Réunion des Musées Nationaux 194; American Museum of Natural History 13, 56; Ancient Art and Architecture Collection Ltd 190, 192, 193b, 196r, 197, 198c, 198r; Rick Matthews/© BBC 2, 8, 16, 22, 33, 36, 47b, 56, 68, 72, 75, 92, 110, 116, 130, 144, 168, 178a, 193a; Alison Brooks 179; Bruce Coleman Collection 62, 104, 131; CASAP/J. Vainstain 149; Hulton Getty 26l; Institute of Human Origins 27l, 27r, 28, 31, 65; John Gurche 84; Javier Trueba/Madrid Scientific Films 118, 119, 121, 122, 132, 134, 135l, 135r, 139; McGregor Museum, South Africa 177; Robin McKie 166; National Geographic 52, 67, 76, 99, 109a, 109b; Natural History Museum, London 20, 49, 54, 57, 126b, 128, 136l, 137c, 137r, 176, 194, 198l; Natural History Museum/Simon Parfitt 126a; Natural History Museum, Netherlands 94, 95a, 95b; National Museums of Kenya/Meave Leakey 24, 25a, 25b, 25c, 47a, 71, 89; Neanderthal Museum, Germany 163; Newsweek International 171; Northern Flagship Institution, South Africa 45l; Peter Pfarr 127; Pictor International 97; Popperfoto 11, 26, 55l, 70; Professor Hillary Deacon/Stellenbosch University 178b; Zev Radovan 208a, 208b; Robert Harding 164; Science Photo Library 12, 15, 30, 34, 38, 40, 42, 44, 45r, 53, 58, 60, 64, 78, 80, 96, 100, 101, 105, 108, 136r, 137l, 140, 146, 148, 180, 181, 183, 195; University of the Witwatersrand, South Africa 41; Alan Walker 73, 74, 82; David Wilson 196l; Joao Zilhao 186

INDEX

Page numbers in *italics* refer to illustrations